Everyday Mathematics®

Student Math Journal 2

The University of Chicago
School Mathematics Project

The McGraw·Hill Companies

UCSMP Elementary Materials Component

Max Bell, Director

Authors

Max Bell
Jean Bell
John Bretzlauf*
Amy Dillard*
Robert Hartfield
Andy Isaacs*
James McBride, Director
Kathleen Pitvorec*
Peter Saecker

Technical Art

Diana Barrie*

**Second Edition only*

Contributors

Carol Arkin, Robert Balfanz, Sharlean Brooks, Ellen Dairyko, James Flanders, David Garcia, Rita Gronbach, Deborah Arron Leslie, Curtis Lieneck, Diana Marnino, Mary Moley, William D. Pattison, William Salvato, Jean Marie Sweigart, Leeann Wille

Photo Credits

Phil Martin/Photography, Jack Demuth/Photography, Cover Credits: Sand, starfish, orange wedges, crystal/Bill Burlingham Photography, Photo Collage: Herman Adler Design Group

Send all inquiries to:
Wright Group/McGraw-Hill
P.O. Box 812960
Chicago, IL 60681

Printed in the United States of America.

ISBN 0-07-584484-2

6 7 8 9 10 11 12 DBH 10 09 08 07 06 05

Contents

Unit 7: Multiplication and Division

A note at the bottom of each journal page indicates when that page is first used. Some pages will be used again during the course of the year.

Unit 8: Fractions

Unit 9: Multiplication and Division

Unit 10: Measurement

Unit 11: Probability; End-of-Year Review

Activity Sheets

Date Time

Product Patterns

Math Message

Complete the facts. Do not use the Multiplication/Division Facts Table.

1. $1 \times 1 =$ ______

2. $2 \times 2 =$ ______

3. $3 \times 3 =$ ______

4. $4 \times 4 =$ ______

5. $5 \times 5 =$ ______

6. $6 \times 6 =$ ______

7. $7 \times 7 =$ ______

8. $8 \times 8 =$ ______

9. $9 \times 9 =$ ______

10. $10 \times 10 =$ ______

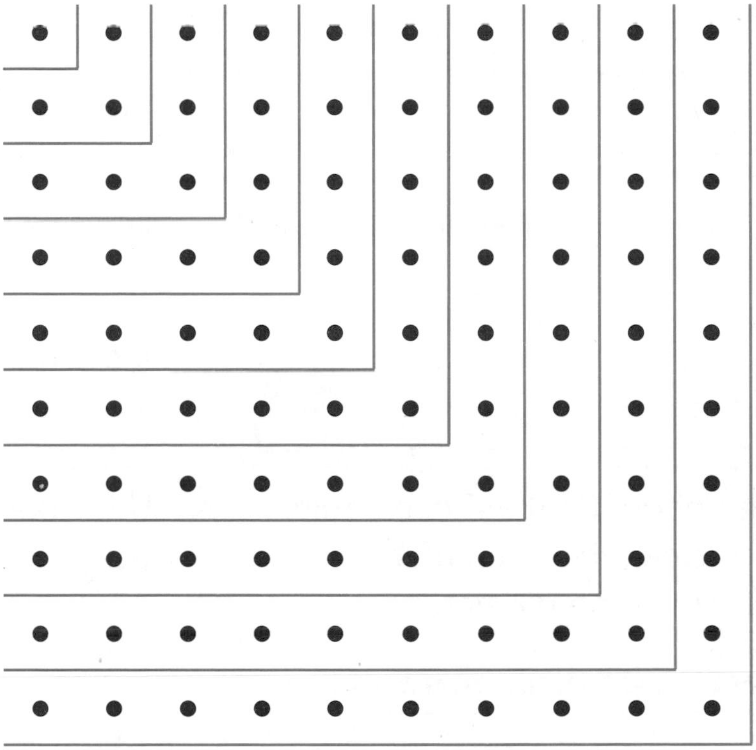

A Two's Product Pattern

Multiply. Look for patterns.

11. $2 \times 2 =$ ______

12. $2 \times 2 \times 2 =$ ______

13. $2 \times 2 \times 2 \times 2 =$ ______

14. $2 \times 2 \times 2 \times 2 \times 2 =$ ______

15. $2 \times 2 \times 2 \times 2 \times 2 \times 2 =$ ______

Challenge

Use the Two's Product Pattern for Problems 11–15. Multiply.

16. $2 \times 2 \times 2 \times 2 \times 2 \times 2 \times 2 =$ ______

Date Time

Math Boxes 7.1

1. Name some items that are shaped like a cone.

____

____

Name some items that are shaped like a cylinder.

____

____

SRB 107

2. In the number 5,627,043:

the 4 means

4 tens

the 6 means

____

the 7 means

____

the 5 means

____

3. Use the "about 3 times" circle rule: For any circle, the circumference is about 3 times the diameter.

diameter	circumference
	24 cm
9 cm	
	21 cm

SRB 134 135

4. Use your calculator.

Enter	Change to	How?
469	1,469	
1,059	859	
23,672	23,972	
46,555	55,555	

SRB 18 19

5. What number is 90 more than 487?

____

What number is 357 less than 608?

____

SRB 51–57

6. 7 baskets. 9 apples per basket. How many apples in all?

__________ apples

8 cakes. 8 candles per cake. How many candles in all?

__________ candles

Date Time

Multiplication/Division Facts Table

×, ÷	1	2	3	4	5	6	7	8	9	10
1	1	2	3	4	5	6	7	8	9	10
2	2	4	6	8	10	12	14	16	18	20
3	3	6	9	12	15	18	21	24	27	30
4	4	8	12	16	20	24	28	32	36	40
5	5	10	15	20	25	30	35	40	45	50
6	6	12	18	24	30	36	42	48	54	60
7	7	14	21	28	35	42	49	56	63	70
8	8	16	24	32	40	48	56	64	72	80
9	9	18	27	36	45	54	63	72	81	90
10	10	20	30	40	50	60	70	80	90	100

Date Time

Math Boxes 7.2

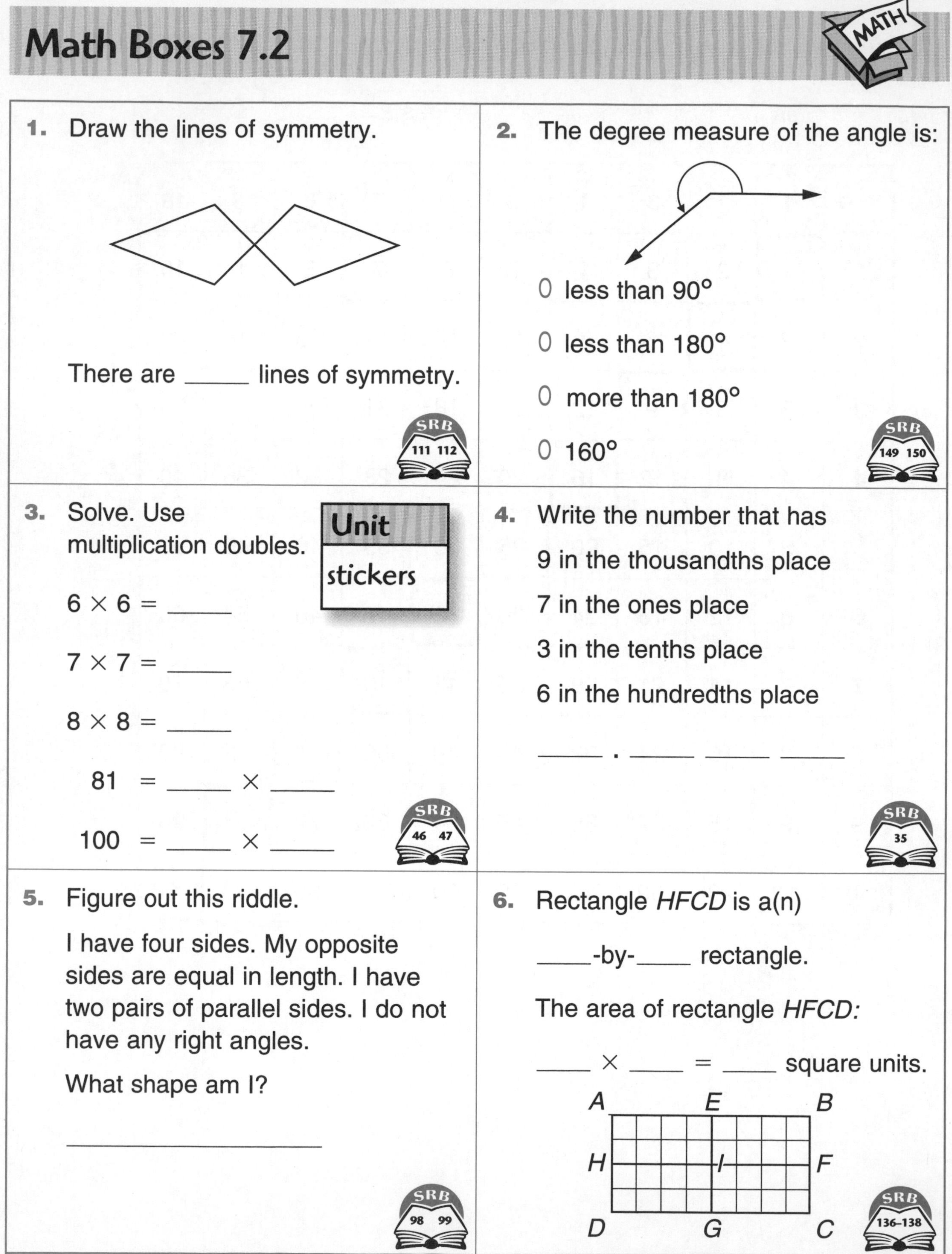

1. Draw the lines of symmetry.

There are _____ lines of symmetry.

SRB 111 112

2. The degree measure of the angle is:

0 less than 90°

0 less than 180°

0 more than 180°

0 160°

SRB 149 150

3. Solve. Use multiplication doubles.

Unit: stickers

$6 \times 6 =$ _____

$7 \times 7 =$ _____

$8 \times 8 =$ _____

$81 =$ _____ $\times$ _____

$100 =$ _____ $\times$ _____

SRB 46 47

4. Write the number that has
9 in the thousandths place
7 in the ones place
3 in the tenths place
6 in the hundredths place

_____ . _____ _____ _____

SRB 35

5. Figure out this riddle.

I have four sides. My opposite sides are equal in length. I have two pairs of parallel sides. I do not have any right angles.

What shape am I?

____

SRB 98 99

6. Rectangle *HFCD* is a(n)

_____-by-_____ rectangle.

The area of rectangle *HFCD:*

_____ $\times$ _____ $=$ _____ square units.

SRB 136–138

Use with Lesson 7.2.

Date Time

Multiplication Bingo

Read the rules for *Multiplication Bingo* on pages 218 and 219 in the *Student Reference Book.*

Write the list of numbers on each grid below.

List of numbers

1	9	18	30
4	12	20	36
6	15	24	50
8	16	25	100

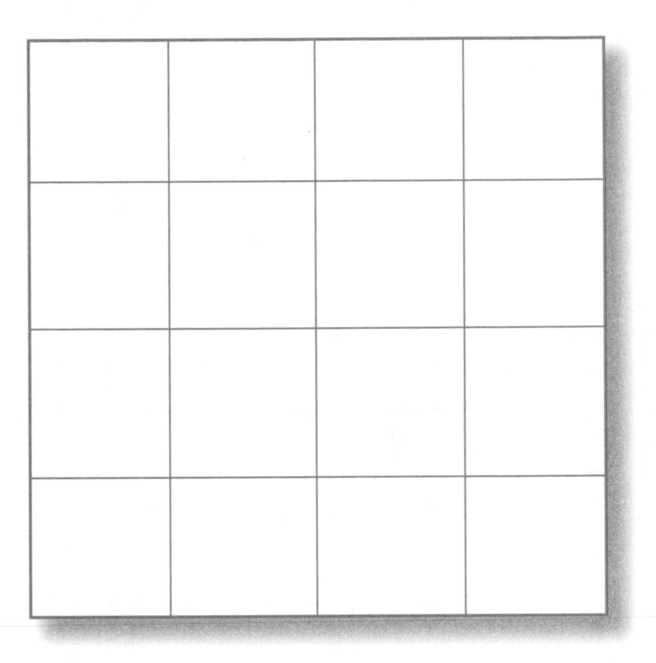

Record the facts you miss.
Practice them in your spare time.

____

____

____

____

____

____

____

Date Time

Multiplication/Division Practice

Fill in the missing number in each Fact Triangle. Then write the fact family for the triangle.

1.

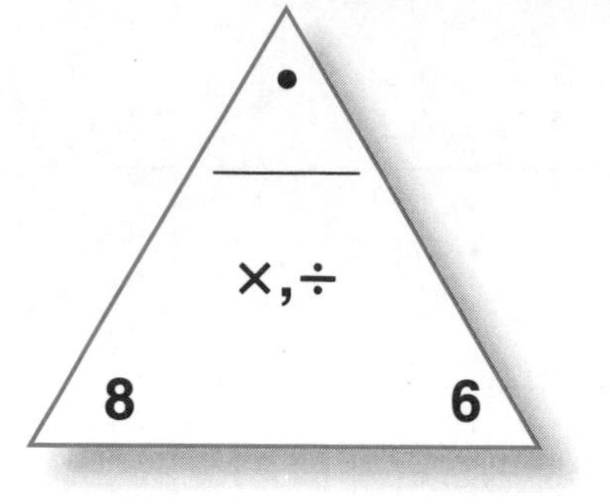

2.

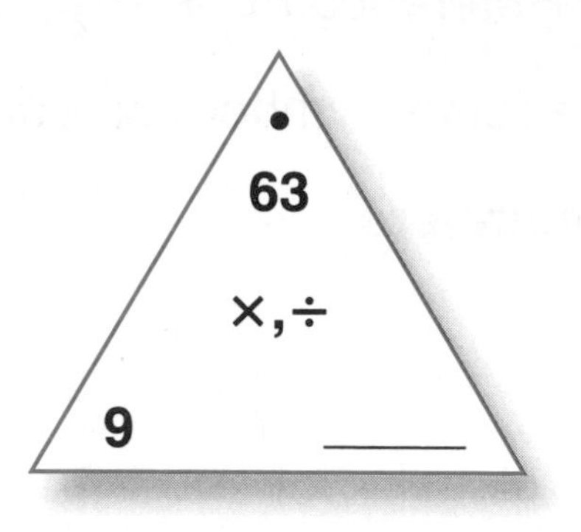

Complete each puzzle.

Example

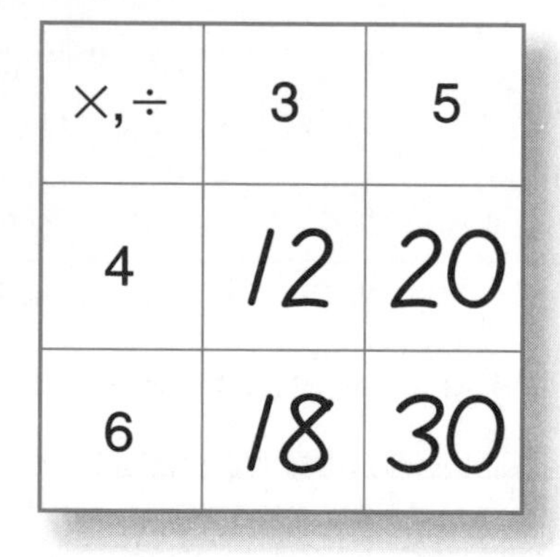

×,÷	3	5
4	12	20
6	18	30

3.

×,÷	2	6
3		
6		

4.

×,÷	3	5
2		
8		

5.

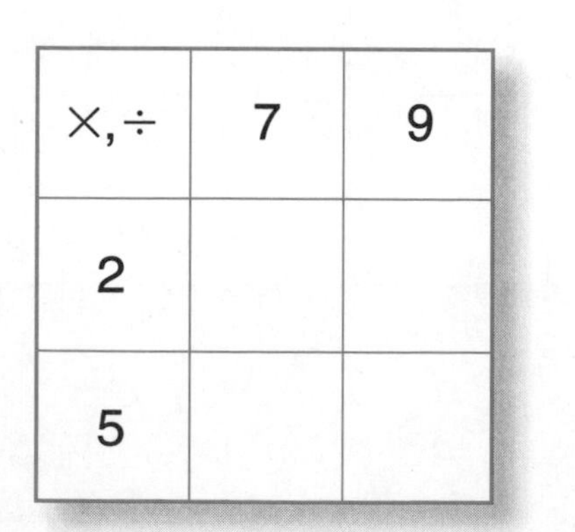

×,÷	7	9
2		
5		

6.

×,÷		4
3	9	
4		

7.

×,÷		6
2		
	24	36

Date Time

Math Boxes 7.3

1. This is a picture of a cube. What do you know about this shape?

____
____
____
____
____

SRB 102–104

2. In the number 5,431,098:

the 3 means thirty thousand

the 4 means _______________

the 9 means _______________

the 5 means _______________

the 1 means _______________

SRB 19

3. Fill in the unit box. Then multiply.

Unit

$8 \times 2 =$ _____

$4 \times 9 =$ _____

$6 \times 7 =$ _____

_____ $= 8 \times 9$

_____ $= 7 \times 8$

SRB 46 47

4. Draw and label a pair of parallel lines. Draw and label a pair of intersecting rays.

SRB 91

5. Add.

$$\begin{array}{r} 349 \\ +\ 956 \\ \hline \end{array} \qquad \begin{array}{r} 777 \\ +\ 1{,}028 \\ \hline \end{array} \qquad \begin{array}{r} 2{,}765 \\ +\ 3{,}842 \\ \hline \end{array}$$

SRB 51–53

6. Maxwell has $806 in the bank. Madison has $589. How much more money does Maxwell have than Madison?

$_______________

SRB 190

Date Time

Number Models with Parentheses

Solve the number story. Then write a number model using parentheses.

1. Amy scored 12 points and Yosh scored 6 points. If their team scored 41 points, how many points did the rest of the team score?

   ____

   Number model: ______________________________

2. In a partner game, Tim has 10 points and Ellen 14 points. They need 50 points to finish the game. How many more points are needed?

   ____

   Number model: ______________________________

3. Once Tim and Ellen got 50 points, but lost 14 points for a wrong move. They gained 10 points back. What was their final score?

   ____

   Number model: ______________________________

Add parentheses to complete the number models.

4. $20 - 10 + 4 = 6$
5. $20 - 10 + 4 = 14$
6. $100 - 21 + 10 = 69$
7. $100 - 21 + 10 = 89$
8. $27 - 8 + 3 = 22$
9. $18 = 6 + 3 \times 4$
10. $5 \times 9 - 2 = 35$
11. $51 = 43 + 15 - 7$

Complete these number models.

12. ______ $= 8 + (9 \times 3)$
13. $(75 - 29) + 5 =$ ______
14. $36 + (15 \div 3) =$ ______
15. ______ $= (8 \times 8) - 16$

Date Time

Math Boxes 7.4

1. Fill in the blanks for this ×, ÷ puzzle.

×, ÷	5	
8		
	45	63

2. Subtract.

$$926 - 538 \qquad 1{,}045 - 471 \qquad 4{,}531 - 2{,}628$$

SRB 54–57

3. Solve.

49 ÷ 7 = _____

81 ÷ 9 = _____

_____ = 64 ÷ 8

6 = 36 ÷ _____

_____ ÷ 5 = 5

SRB 46 47

4. 1st grade collected 545 pop cans.
2nd grade collected 766 pop cans.
3rd grade collected 802 pop cans.

How many in all?

__________ pop cans

SRB 188 189

5. Draw a parallelogram. Label the vertices so that $\overline{AB} \parallel \overline{CD}$. The symbol ∥ means "is parallel to."

SRB 98 99

6. How many children like green? _____ children
How many children in all responded to the question? _____ children

Favorite Color	Number of Children
blue	~~||||~~ ~~||||~~
red	~~||||~~ |||
green	~~||||~~ ~~||||~~ |
other	~~||||~~ ~~||||~~ ||||

SRB 70–72

Date Time

Scoring 10 Basketball Points

Find different ways to score 10 points in a basketball game.

Number of 3-point baskets	Number of 2-point baskets	Number of 1-point baskets	Number models
2	2	0	$(2 \times 3) + (2 \times 2) + (0 \times 1) = 10$

Use with Lesson 7.5.

Date Time

Multiplication and Division Practice

Fill in the missing number in each Fact Triangle.
Then write the fact family for the triangle.

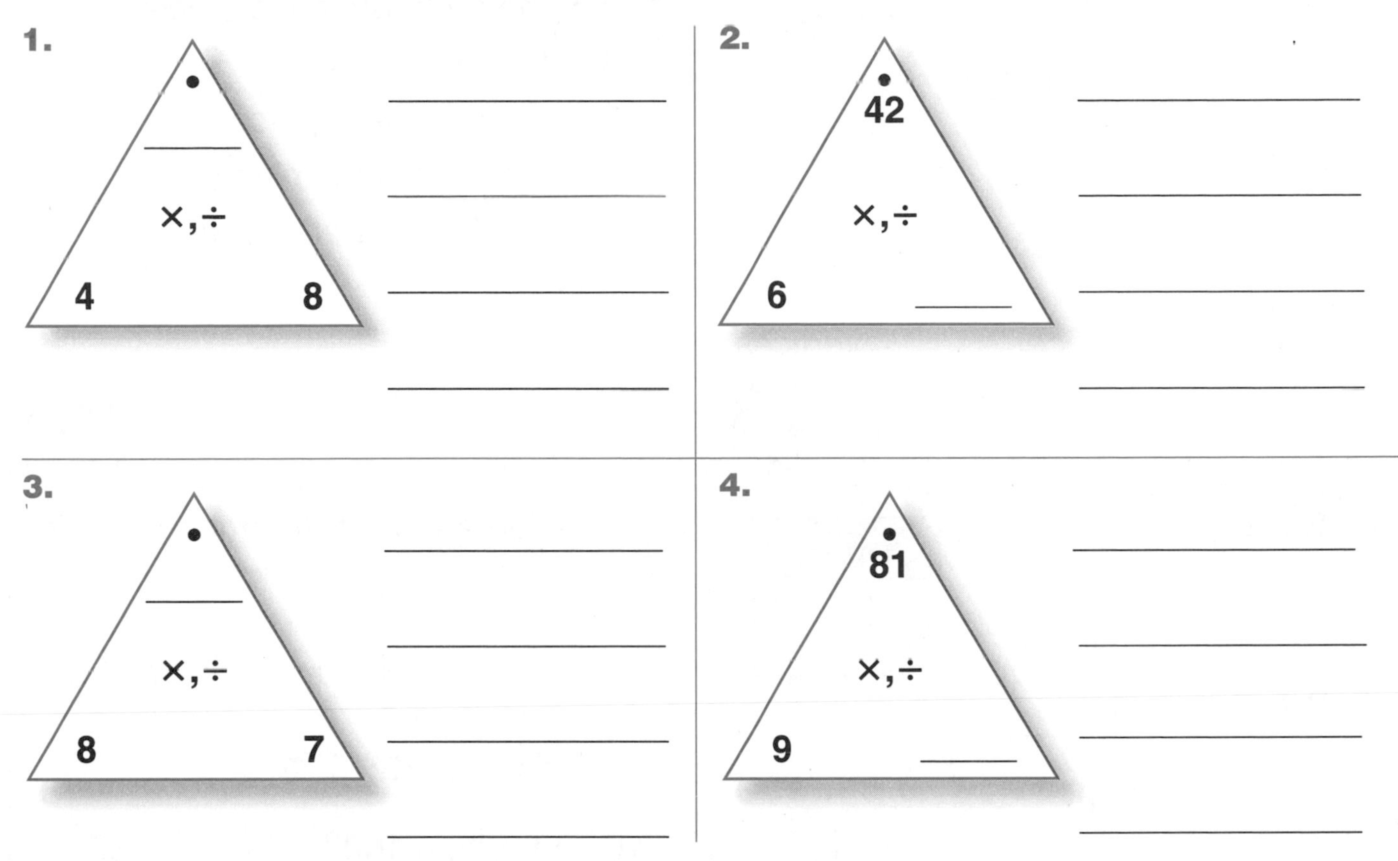

5. Circle the Fact Triangle above that shows a square product.

Fill in the tables and find the missing rule.

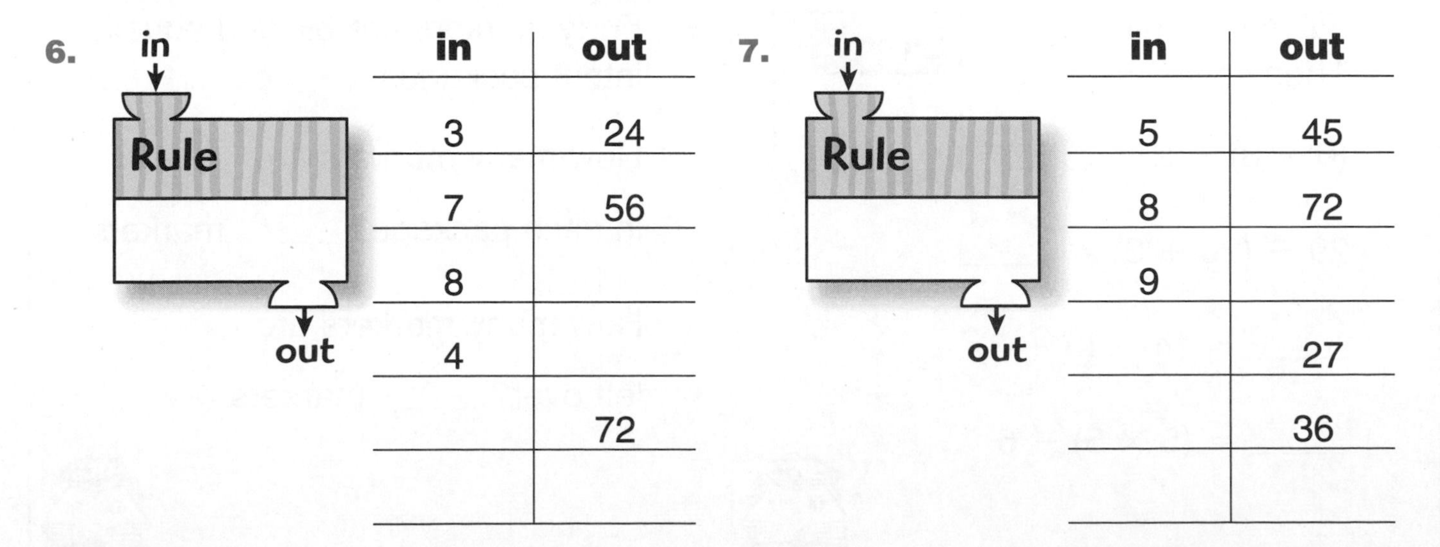

6.

in	out
3	24
7	56
8	
4	
	72

7.

in	out
5	45
8	72
9	
	27
	36

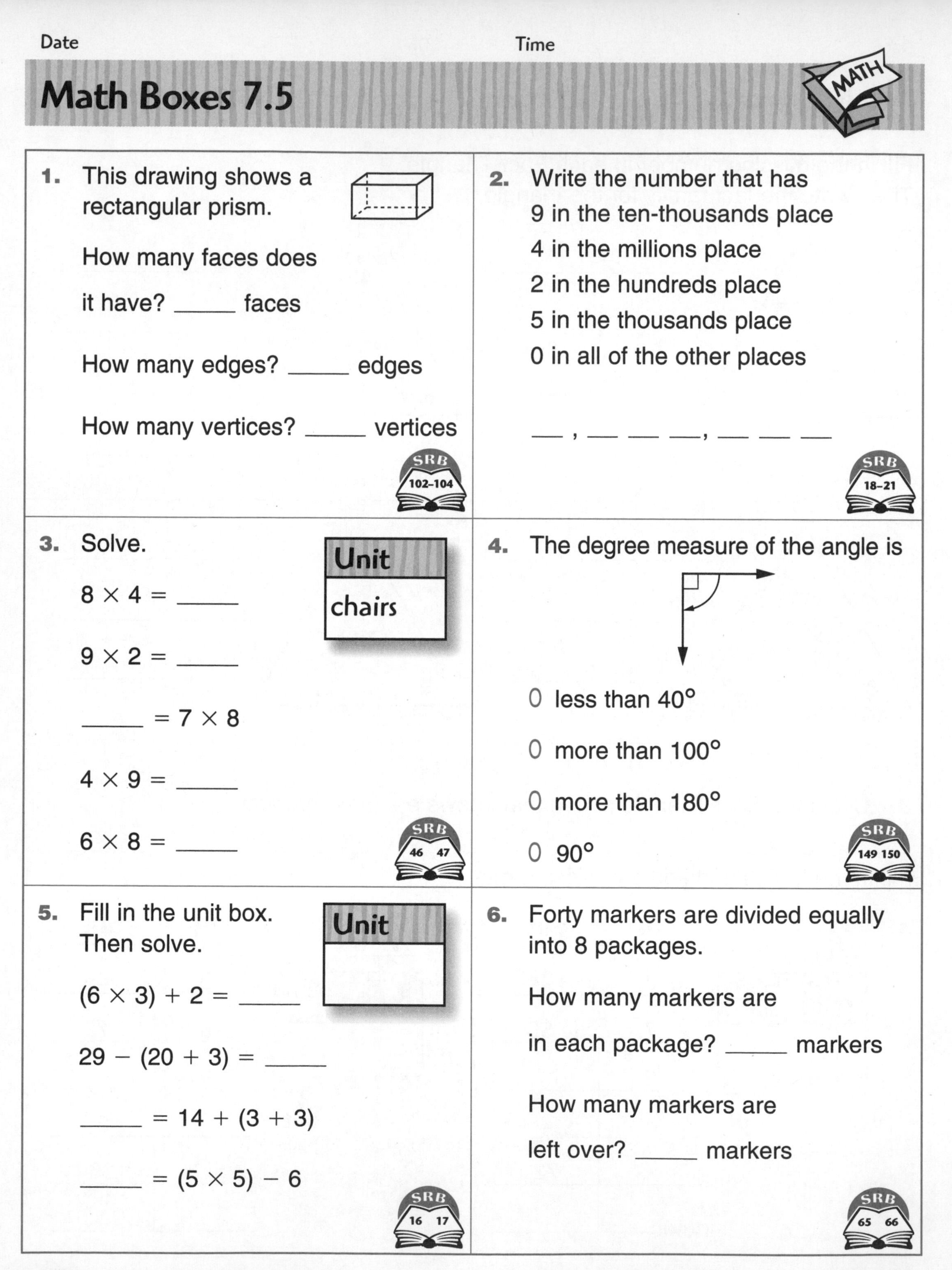

Date Time

Math Boxes 7.5

1. This drawing shows a rectangular prism.

How many faces does it have? _____ faces

How many edges? _____ edges

How many vertices? _____ vertices

SRB 102–104

2. Write the number that has
9 in the ten-thousands place
4 in the millions place
2 in the hundreds place
5 in the thousands place
0 in all of the other places

___ , ___ ___ ___, ___ ___ ___

SRB 18–21

3. Solve.

Unit
chairs

$8 \times 4 =$ _____

$9 \times 2 =$ _____

_____ $= 7 \times 8$

$4 \times 9 =$ _____

$6 \times 8 =$ _____

SRB 46 47

4. The degree measure of the angle is

0 less than 40°

0 more than 100°

0 more than 180°

0 90°

SRB 149 150

5. Fill in the unit box. Then solve.

Unit

$(6 \times 3) + 2 =$ _____

$29 - (20 + 3) =$ _____

_____ $= 14 + (3 + 3)$

_____ $= (5 \times 5) - 6$

SRB 16 17

6. Forty markers are divided equally into 8 packages.

How many markers are in each package? _____ markers

How many markers are left over? _____ markers

SRB 65 66

Date Time

Extended Multiplication and Division Facts

Write the number of 3s in each number.

1. How many 3s in 30? ________
2. How many 3s in 300? ________
3. How many 3s in 3,000? ________
4. How many 3s in 12? ________
5. How many 3s in 120? ________
6. How many 3s in 1,200? ________

Solve each ×, ÷ puzzle. Fill in the blanks.

Example

×, ÷	300	2,000
2	600	4,000
3	900	6,000

7.

×, ÷	60	300
4		
5	300	

8.

×, ÷	4	5
200		
8,000		

9.

×, ÷		1,000
3	1,500	
		6,000

Solve each number story.

10. A 30-minute television program has two 60-second commercials at the beginning and two at the end. There are also four 30-second commercials in the middle of the program. How long is the actual program?

    ________ minutes

11. During a 40-minute basketball game, each team is allowed four 60-second timeouts and two 30-second timeouts. If both teams use all of their timeouts, how many minutes of timeouts will there be?

    ________ minutes

Math Boxes 7.6

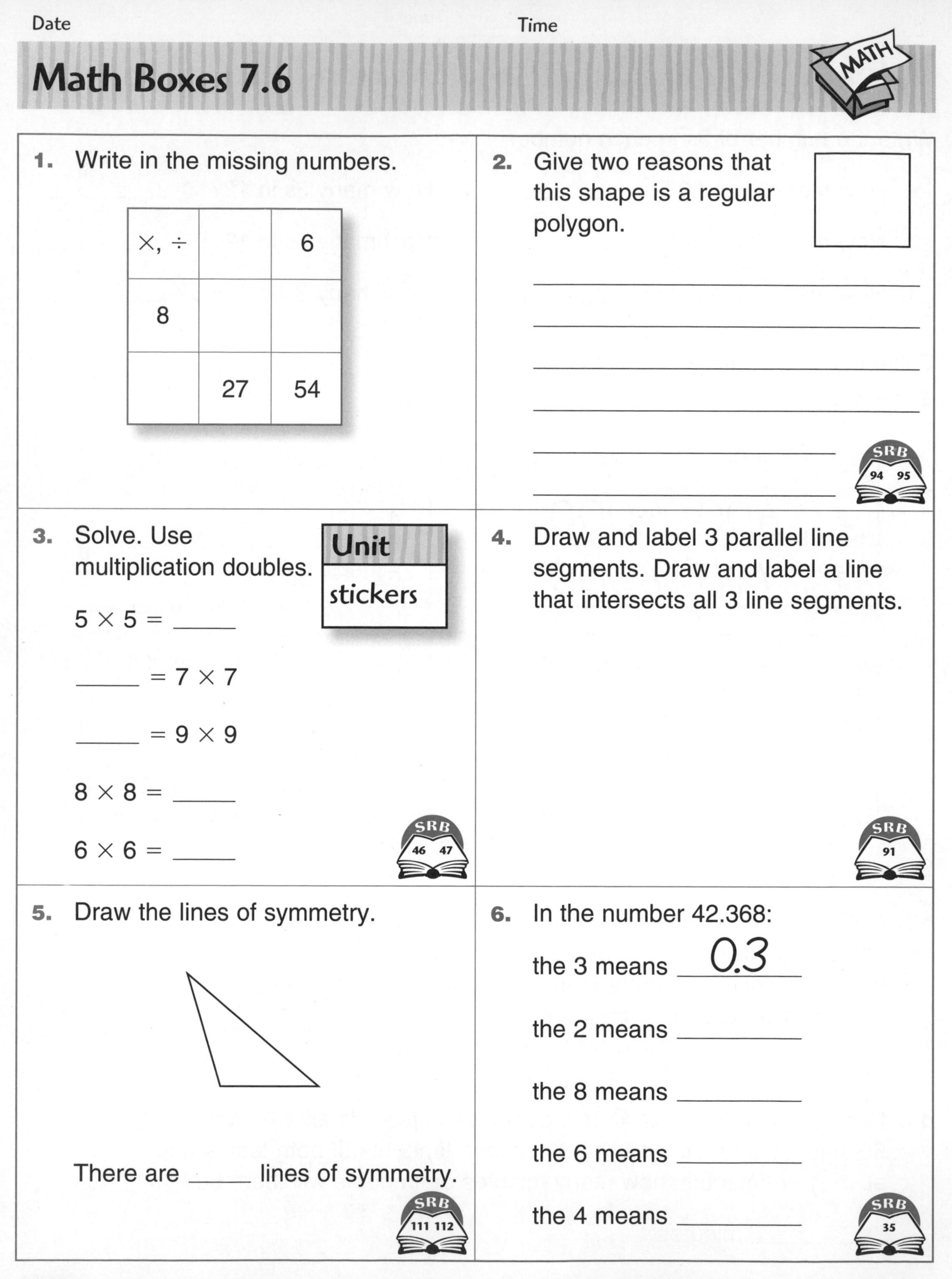

1. Write in the missing numbers.

×, ÷		6
8		
	27	54

2. Give two reasons that this shape is a regular polygon.

____

____

____

____

____

____

SRB 94 95

3. Solve. Use multiplication doubles.

Unit
stickers

$5 \times 5 =$ _____

_____ $= 7 \times 7$

_____ $= 9 \times 9$

$8 \times 8 =$ _____

$6 \times 6 =$ _____

SRB 46 47

4. Draw and label 3 parallel line segments. Draw and label a line that intersects all 3 line segments.

SRB 91

5. Draw the lines of symmetry.

There are _____ lines of symmetry.

SRB 111 112

6. In the number 42.368:

the 3 means 0.3

the 2 means __________

the 8 means __________

the 6 means __________

the 4 means __________

SRB 35

Date Time

Stock-Up Sale Record

Use the items on pages 240 and 241 in your *Student Reference Book.*

Round 1:

Item to be purchased: ____________

How many? ____________________

Regular or sale price? ____________

Price per item: __________________

Estimated cost: _________________

Round 2:

Item to be purchased: ____________

How many? ____________________

Regular or sale price? ____________

Price per item: __________________

Estimated cost: _________________

Round 3:

Item to be purchased: ____________

How many? ____________________

Regular or sale price? ____________

Price per item: __________________

Estimated cost: _________________

Round 4:

Item to be purchased: ____________

How many? ____________________

Regular or sale price? ____________

Price per item: __________________

Estimated cost: _________________

Round 5:

Item to be purchased: ____________

How many? ____________________

Regular or sale price? ____________

Price per item: __________________

Estimated cost: _________________

Round 6:

Item to be purchased: ____________

How many? ____________________

Regular or sale price? ____________

Price per item: __________________

Estimated cost: _________________

Date Time

Extended Facts Practice

Solve the calculator puzzles. Use the (×) or (÷) for each puzzle.

	Enter	Change to	How?
1.	10	1,000	______
2.	1,000	100	______
3.	100	10,000	______
4.	1,000	10	______
5.	10,000	1,000	______

6. Three of the names do not belong in this name-collection box. Cross them out.

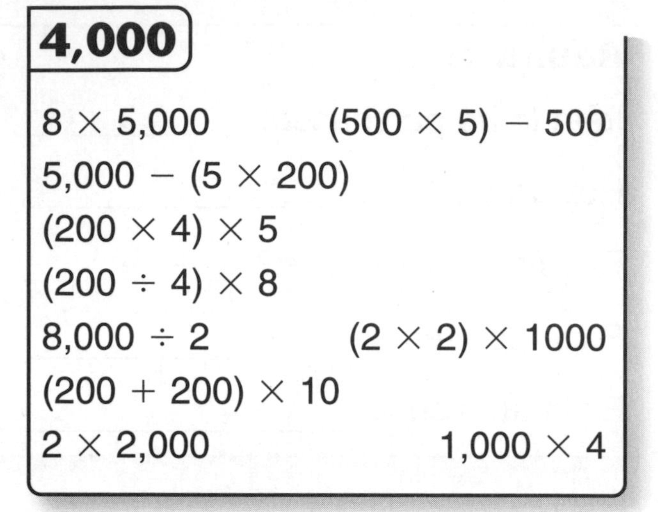

4,000
8 × 5,000
(500 × 5) − 500
5,000 − (5 × 200)
(200 × 4) × 5
(200 ÷ 4) × 8
8,000 ÷ 2
(2 × 2) × 1000
(200 + 200) × 10
2 × 2,000
1,000 × 4

Complete the extended Fact Triangles. Write the extended fact families.

7.

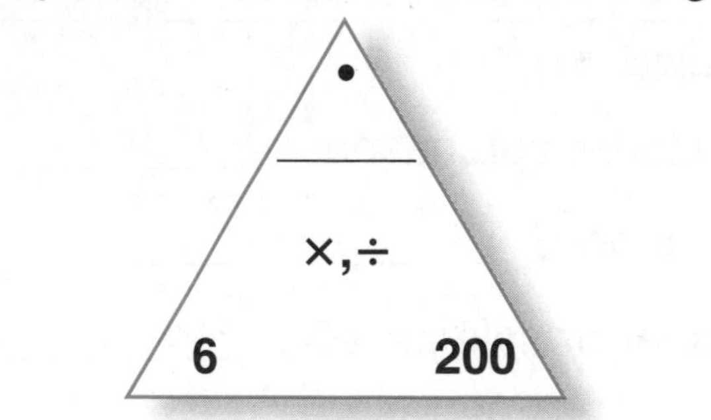

____

____

____

____

8.

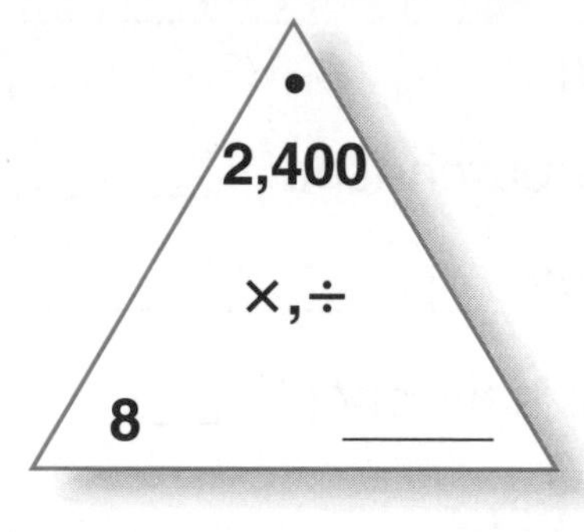

____

____

____

____

Solve each ×, ÷ puzzle. Fill in the blanks.

9.

×, ÷	3	9
100		900
3,000		

10.

×, ÷		2,000
4	1,200	
		10,000

Date Time

Math Boxes 7.7

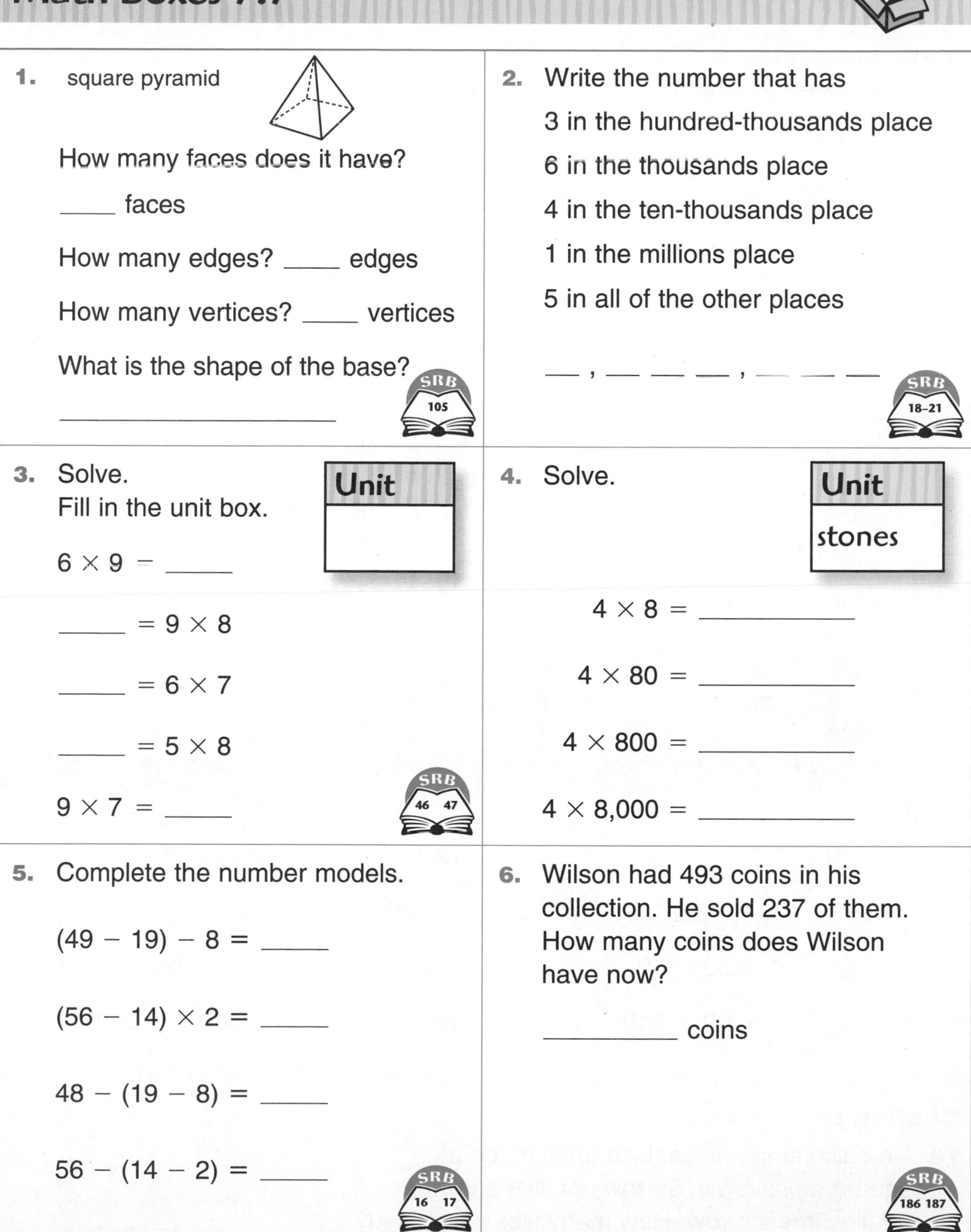

1. square pyramid

How many faces does it have?

_____ faces

How many edges? _____ edges

How many vertices? _____ vertices

What is the shape of the base?

____

SRB 105

2. Write the number that has

3 in the hundred-thousands place

6 in the thousands place

4 in the ten-thousands place

1 in the millions place

5 in all of the other places

___ , ___ ___ ___ , ___ ___ ___

SRB 18–21

3. Solve.
Fill in the unit box.

Unit

$6 \times 9 =$ _____

_____ $= 9 \times 8$

_____ $= 6 \times 7$

_____ $= 5 \times 8$

$9 \times 7 =$ _____

SRB 46 47

4. Solve.

Unit
stones

$4 \times 8 =$ _____

$4 \times 80 =$ _____

$4 \times 800 =$ _____

$4 \times 8{,}000 =$ _____

5. Complete the number models.

$(49 - 19) - 8 =$ _____

$(56 - 14) \times 2 =$ _____

$48 - (19 - 8) =$ _____

$56 - (14 - 2) =$ _____

SRB 16 17

6. Wilson had 493 coins in his collection. He sold 237 of them. How many coins does Wilson have now?

_________ coins

SRB 186 187

Date Time

Tens Times Tens

Math Message

Write the dollar values.

1. 10 [$10] = $____________

2. 100 [$10] = $____________

3. 1,000 [$10] = $____________

4. 10 [$100] = $____________

5. 100 [$100] = $____________

6. 1,000 [$100] = $____________

Solve each ×, ÷ puzzle. Fill in the blanks.

7.

×, ÷	10	100
1		
10		

8.

×, ÷	4	30
20		
6		

9.

×, ÷	40	60
20		
80		

10.

×, ÷		
3	150	
70		560

Multiply.

11. 5 × 90 = ____________

12. ____________ = 70 × 4

13. 10 × 70 = ____________

14. 80 × 60 = ____________

15. ____________ = 30 × 50

16. 7 × ____________ = 420

17. ____________ × 90 = 540

18. ____________ × 600 = 6,000

Challenge

19. No calculators, please! An artist made a square mosaic with 99 rows of tiles and 99 tiles in each row. How many tiles were used? ____________ (unit)

Date Time

Math Boxes 7.8

1. Fill in the blanks for this ×,÷ puzzle.

×,÷	9	6
7		
9		

2. Multiply.

5×9 50×9 500×9

3×9 30×9 300×9

3. Write multiplication names for three different square numbers.

____

____

____

4. The best estimate of 5,697 + 1,310 is:

0 about 8,100

0 about 8,000

0 about 7,000

0 about 5,901

SRB 168

5. Add parentheses to complete the number models.

30 = 10 × 2 + 10

46 − 23 − 13 = 10

4 ÷ 2 + 6 = 8

SRB 16 17

6. Draw an angle that measures between 90° and 180°.

SRB 149 150

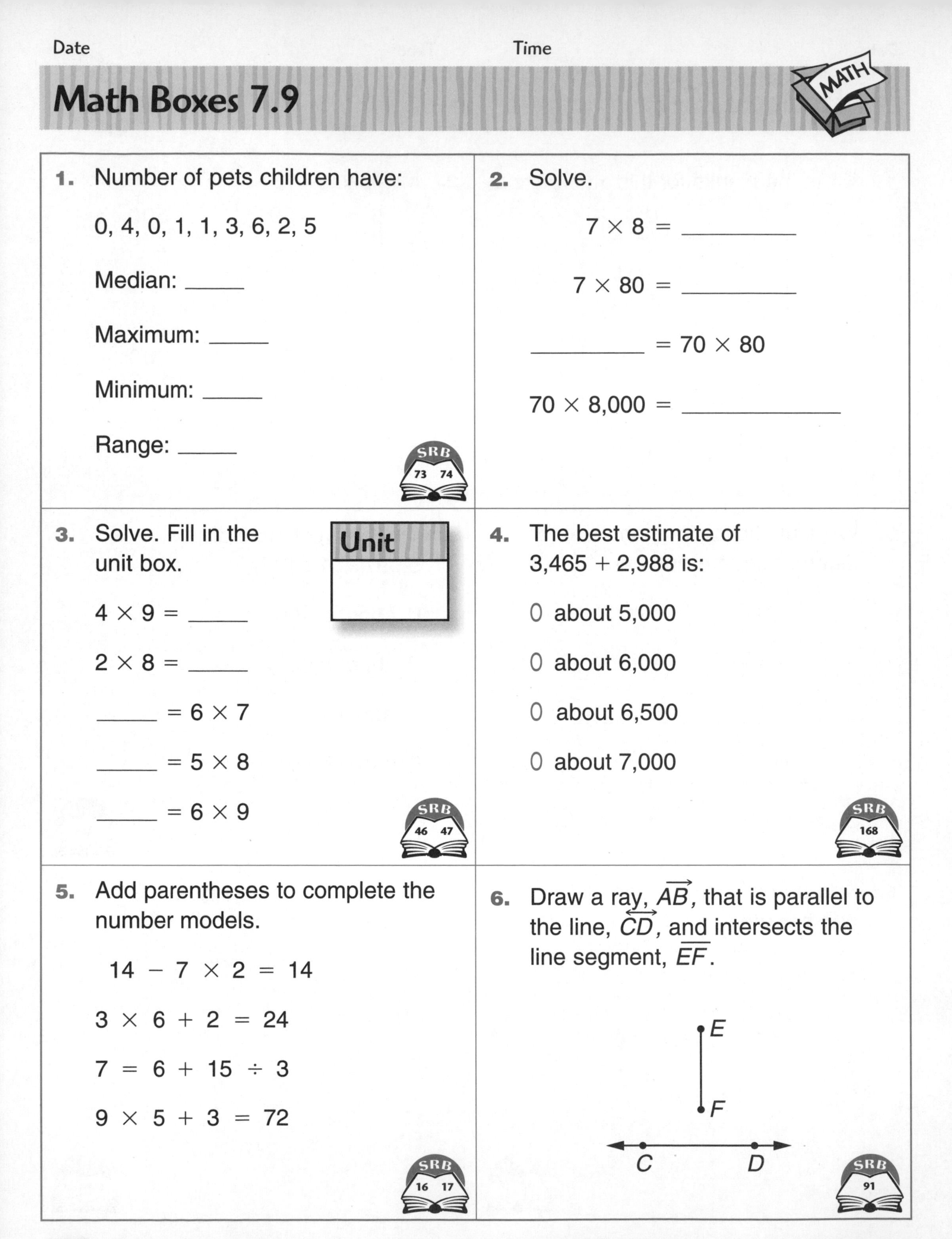

Date Time

Math Boxes 7.9

1. Number of pets children have:

0, 4, 0, 1, 1, 3, 6, 2, 5

Median: _____

Maximum: _____

Minimum: _____

Range: _____

2. Solve.

$7 \times 8 =$ __________

$7 \times 80 =$ __________

__________ $= 70 \times 80$

$70 \times 8{,}000 =$ ______________

3. Solve. Fill in the unit box.

Unit

$4 \times 9 =$ _____

$2 \times 8 =$ _____

_____ $= 6 \times 7$

_____ $= 5 \times 8$

_____ $= 6 \times 9$

4. The best estimate of $3{,}465 + 2{,}988$ is:

0 about 5,000

0 about 6,000

0 about 6,500

0 about 7,000

5. Add parentheses to complete the number models.

$14 - 7 \times 2 = 14$

$3 \times 6 + 2 = 24$

$7 = 6 + 15 \div 3$

$9 \times 5 + 3 = 72$

6. Draw a ray, $\overrightarrow{AB}$, that is parallel to the line, $\overleftrightarrow{CD}$, and intersects the line segment, $\overline{EF}$.

Date Time

Math Boxes 7.10

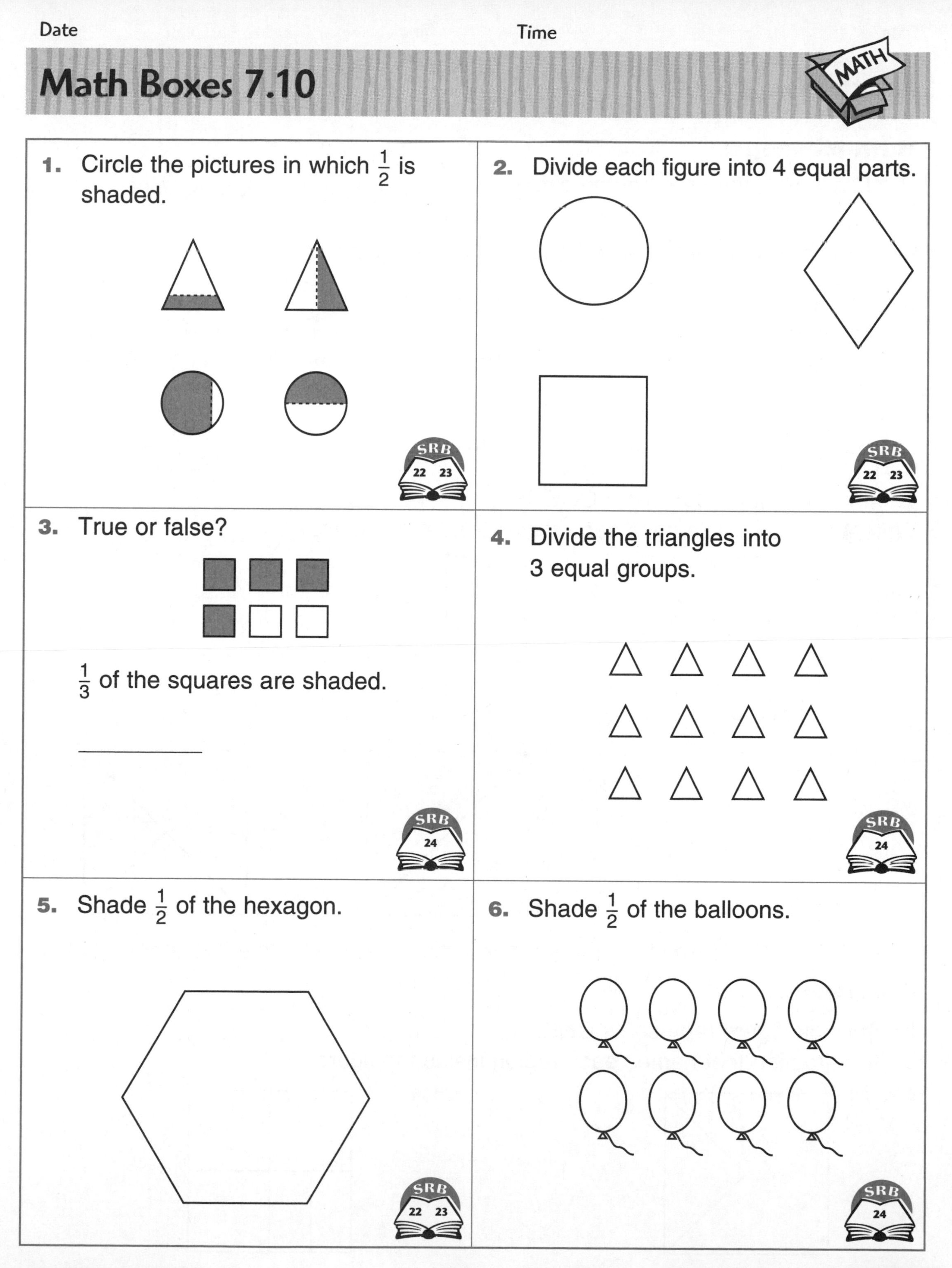

1. Circle the pictures in which $\frac{1}{2}$ is shaded.

2. Divide each figure into 4 equal parts.

3. True or false?

   $\frac{1}{3}$ of the squares are shaded.

   ______

4. Divide the triangles into 3 equal groups.

5. Shade $\frac{1}{2}$ of the hexagon.

6. Shade $\frac{1}{2}$ of the balloons.

Date Time

Fraction Review

Math Message

1. Draw an X through $\frac{2}{3}$ of the circles.

Label each picture with one of the following numbers: 0, $\frac{0}{4}$, $\frac{1}{4}$, $\frac{1}{2}$, $\frac{2}{4}$, or $\frac{3}{4}$.

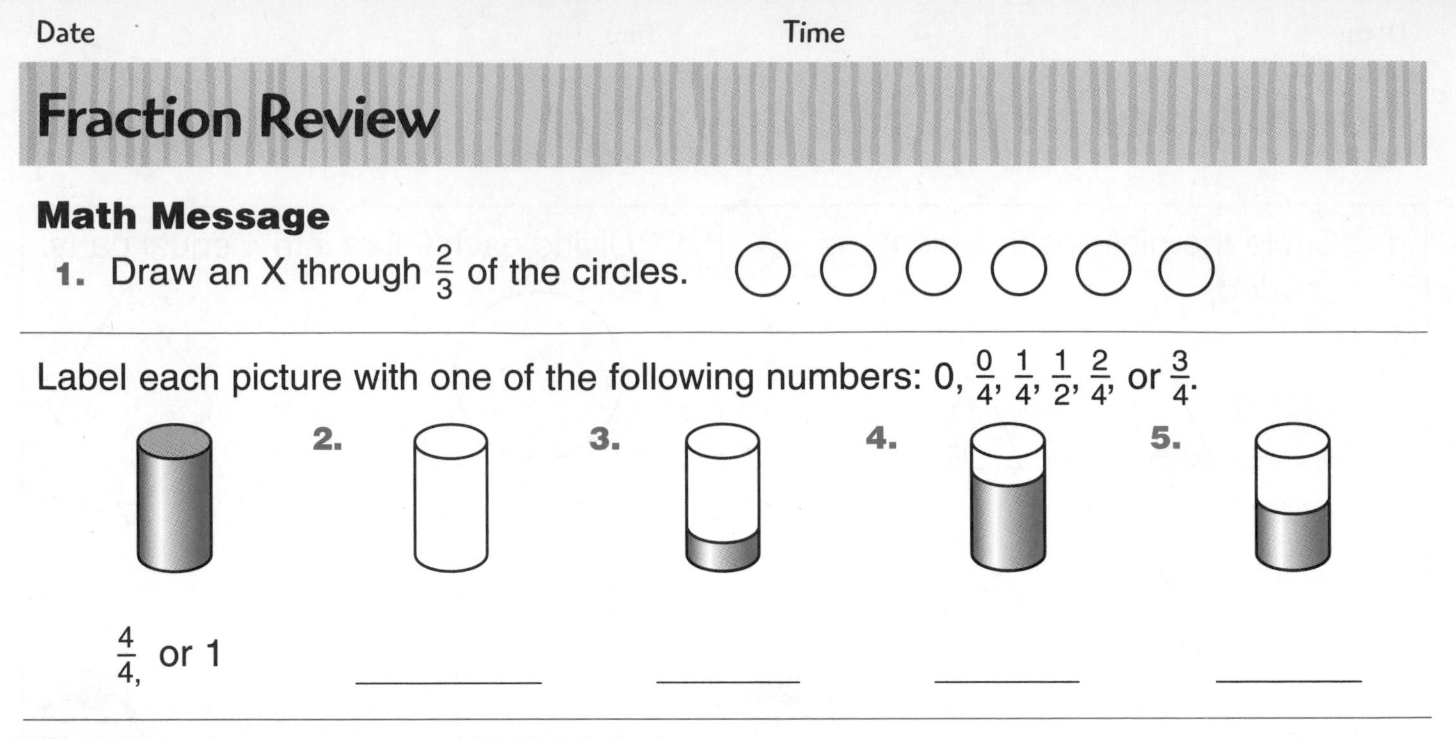

Each whole figure represents ONE.
Write a fraction that names each region inside the figure.

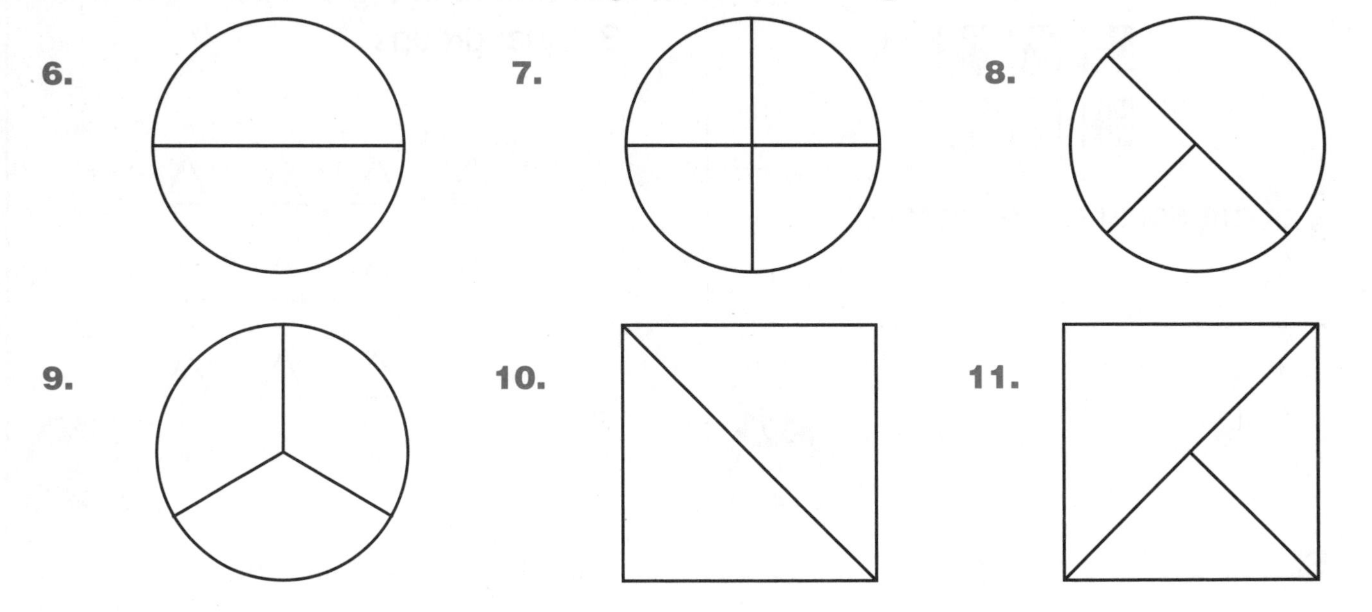

Challenge

Each whole figure represents ONE.
Write a fraction that names each region inside the figure.

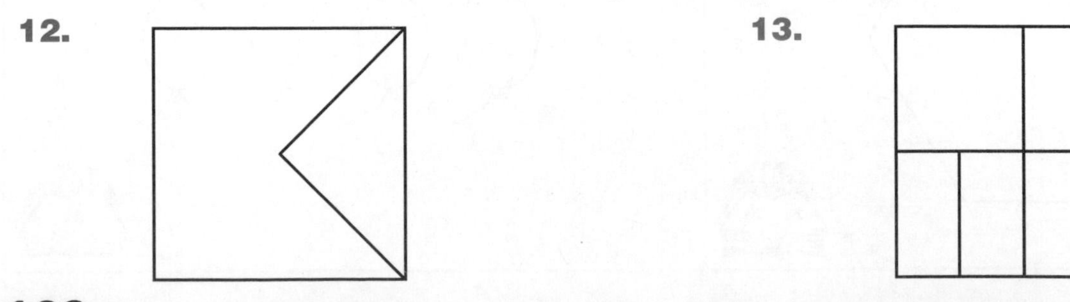

Date Time

Fraction Review (cont.)

You need at least 25 pennies or other counters. Use them to help you solve these problems. Share solution strategies with others in your group.

Unit
counters

Make a set of 8 counters.

14. Show $\frac{1}{4}$ of a set of 8 counters. How many counters is that? ______

15. Put the counters back. Show $\frac{2}{4}$ of the set. How many counters? ______

16. Put the counters back. Show $\frac{3}{4}$ of the set. How many counters? ______

Make a set of 12 counters.

17. Show $\frac{1}{3}$ of the set. How many counters is that? ______

18. Put the counters back. Show $\frac{2}{3}$ of the set. How many counters? ______

19. Put the counters back. Show $\frac{3}{3}$ of the set. How many counters? ______

20. Show $\frac{1}{5}$ of a set of 15 counters. How many counters is that? ______

21. Show $\frac{4}{5}$ of a set of 15 counters. How many counters is that? ______

22. Show $\frac{3}{4}$ of a set of 20 counters. How many counters is that? ______

23. Show $\frac{2}{3}$ of a set of 18 counters. How many counters? ______

24. Five counters is $\frac{1}{5}$ of a set. How many are in the whole set? ______

25. Six counters is $\frac{1}{3}$ of a set. How many are in the whole set? ______

Challenge

26. Twelve counters is $\frac{3}{4}$ of a set. How many are in the complete set of counters? ______

27. Pretend that you have 15 "cheese cubes" that can be cut. How many are in $\frac{1}{2}$ of the set of cubes? Use a fraction or decimal in your answer. ______

Math Boxes 8.1

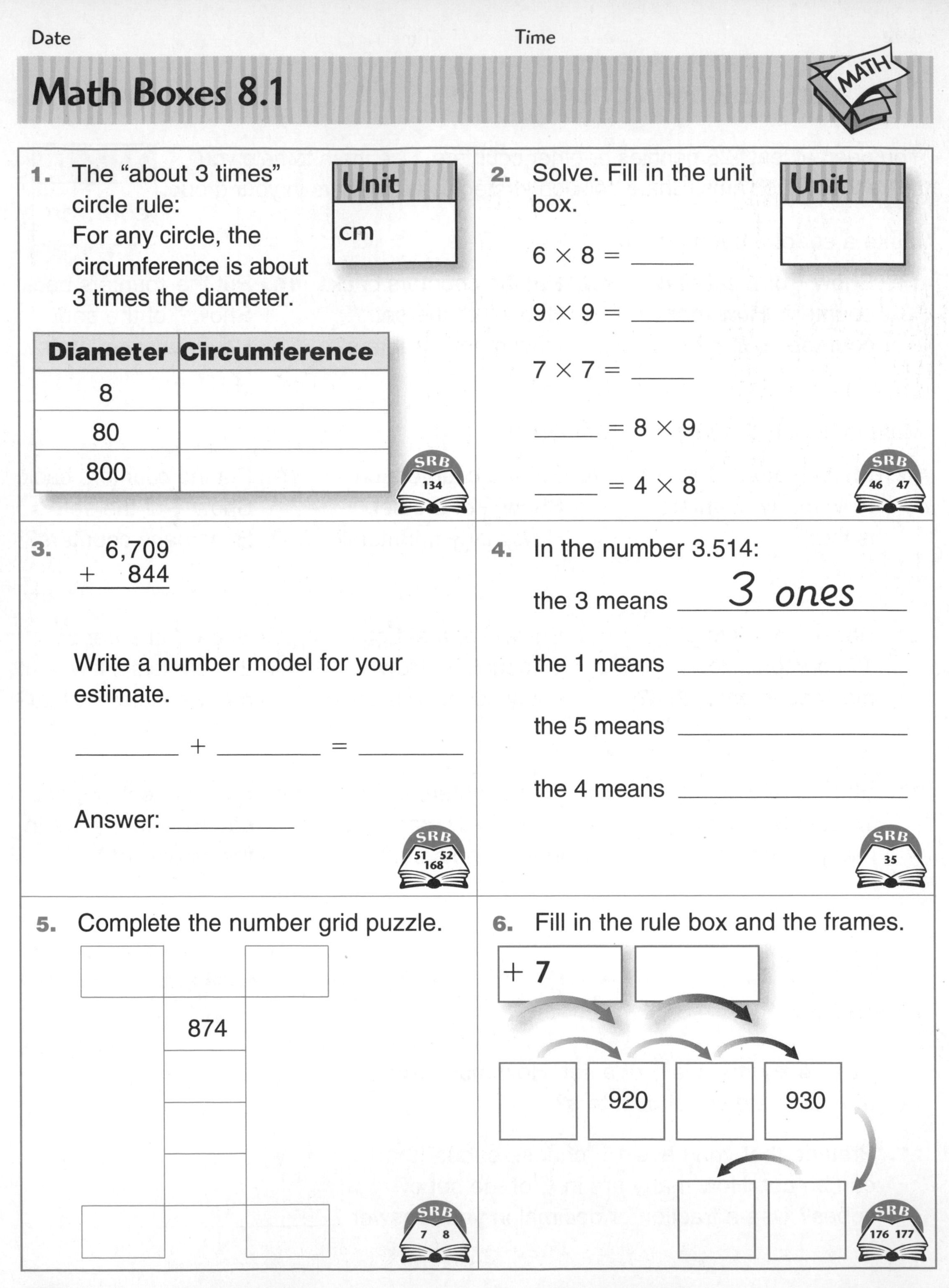

1. The "about 3 times" circle rule:
For any circle, the circumference is about 3 times the diameter.

Unit
cm

Diameter	Circumference
8	
80	
800	

SRB 134

2. Solve. Fill in the unit box.

Unit

$6 \times 8 =$ _____

$9 \times 9 =$ _____

$7 \times 7 =$ _____

_____ $= 8 \times 9$

_____ $= 4 \times 8$

SRB 46 47

3. 6,709
+ 844

Write a number model for your estimate.

_______ + _______ = _______

Answer: _______

SRB 51 52 168

4. In the number 3.514:

the 3 means 3 ones

the 1 means _______

the 5 means _______

the 4 means _______

SRB 35

5. Complete the number grid puzzle.

874

SRB 7 8

6. Fill in the rule box and the frames.

+ 7

920

930

SRB 176 177

Date Time

Fractions with Pattern Blocks

Work with a partner.

Materials
- ❑ pattern blocks
- ❑ Pattern-Block Template

Part 1

Cover each shape with green △ pattern blocks. What fractional part of each shape is 1 green pattern block? Write the fraction under each shape.

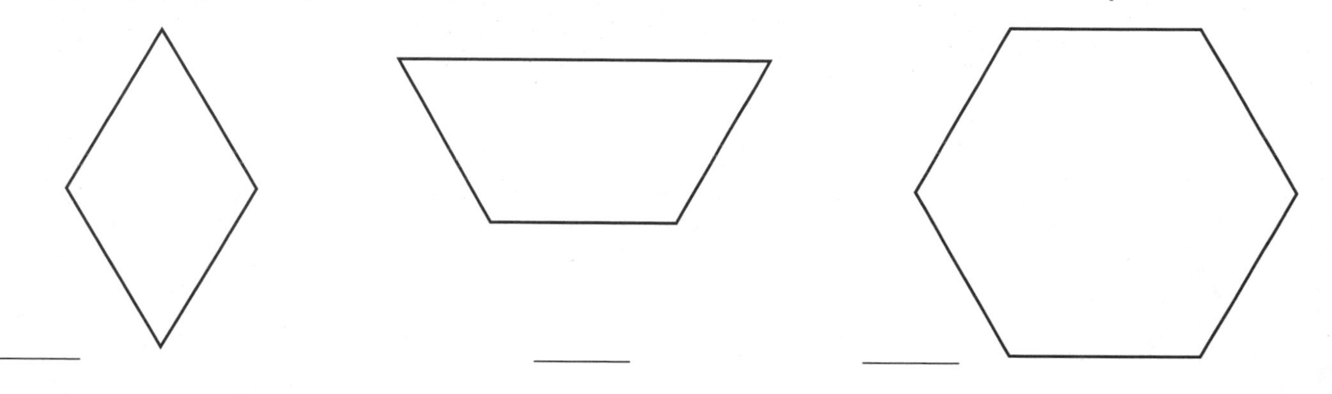

Part 2

Cover each shape with green △ pattern blocks. What fractional part of each shape are 2 green pattern blocks? Write the fraction next to each shape.

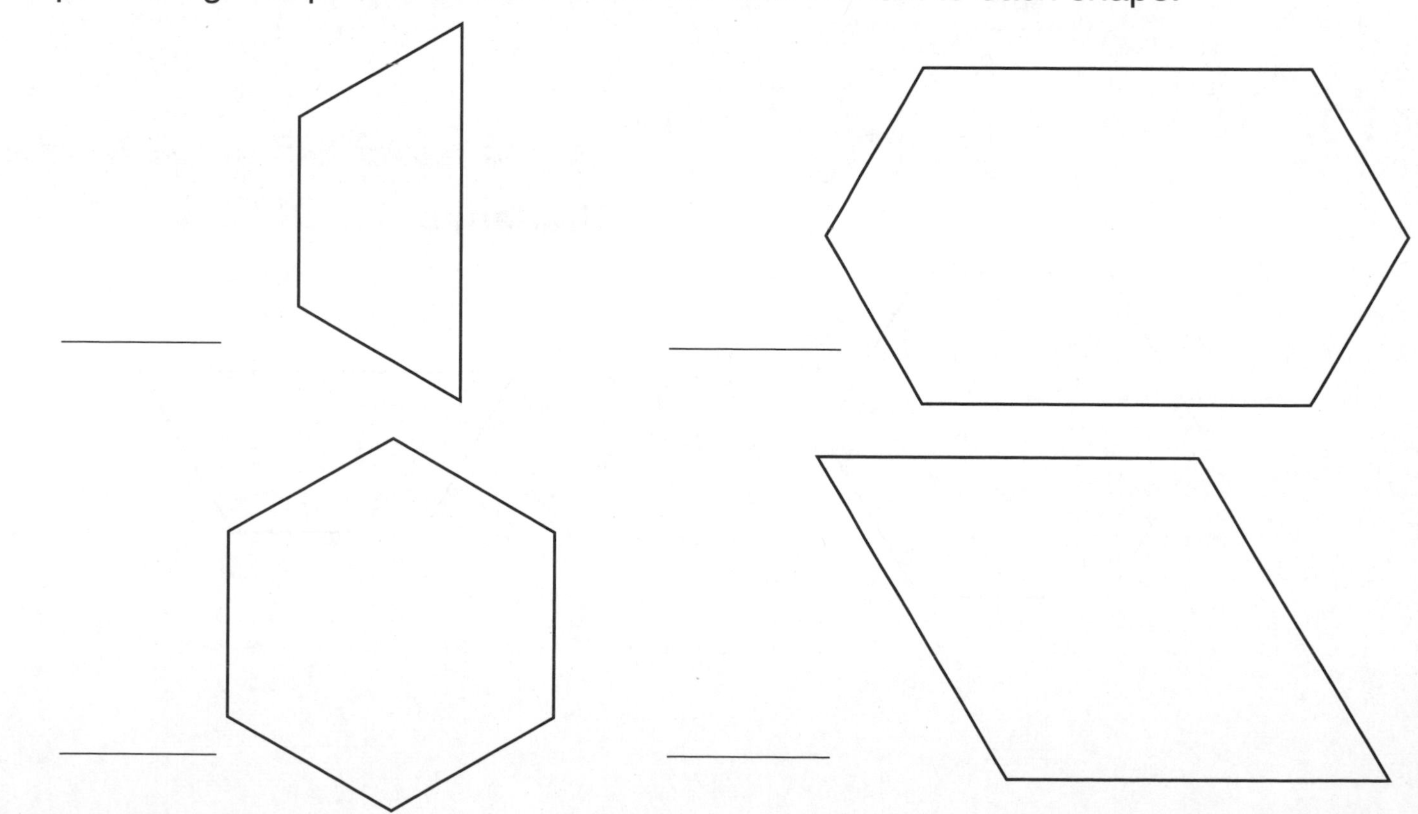

Date Time

Fractions with Pattern Blocks (cont.)

Part 3

Cover each shape with blue ◇ pattern blocks. What fractional part of each shape is 1 blue pattern block? Write the fraction under each shape. If you can't cover the whole shape, cover as much as you can.
Think: Is there another block that would cover the rest of the shape?

______

______

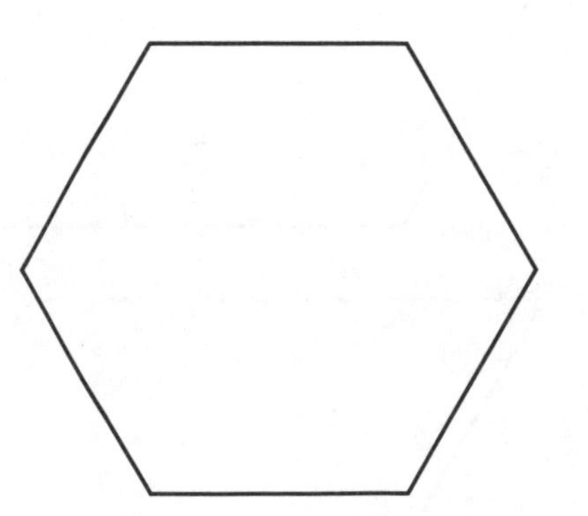

______

Challenge

______

Date Time

Fractions with Pattern Blocks (cont.)

Part 4

Cover each shape with blue ◇ pattern blocks. What fractional part of each shape would 2 blue pattern blocks cover? Write the fraction next to each shape.

Challenge

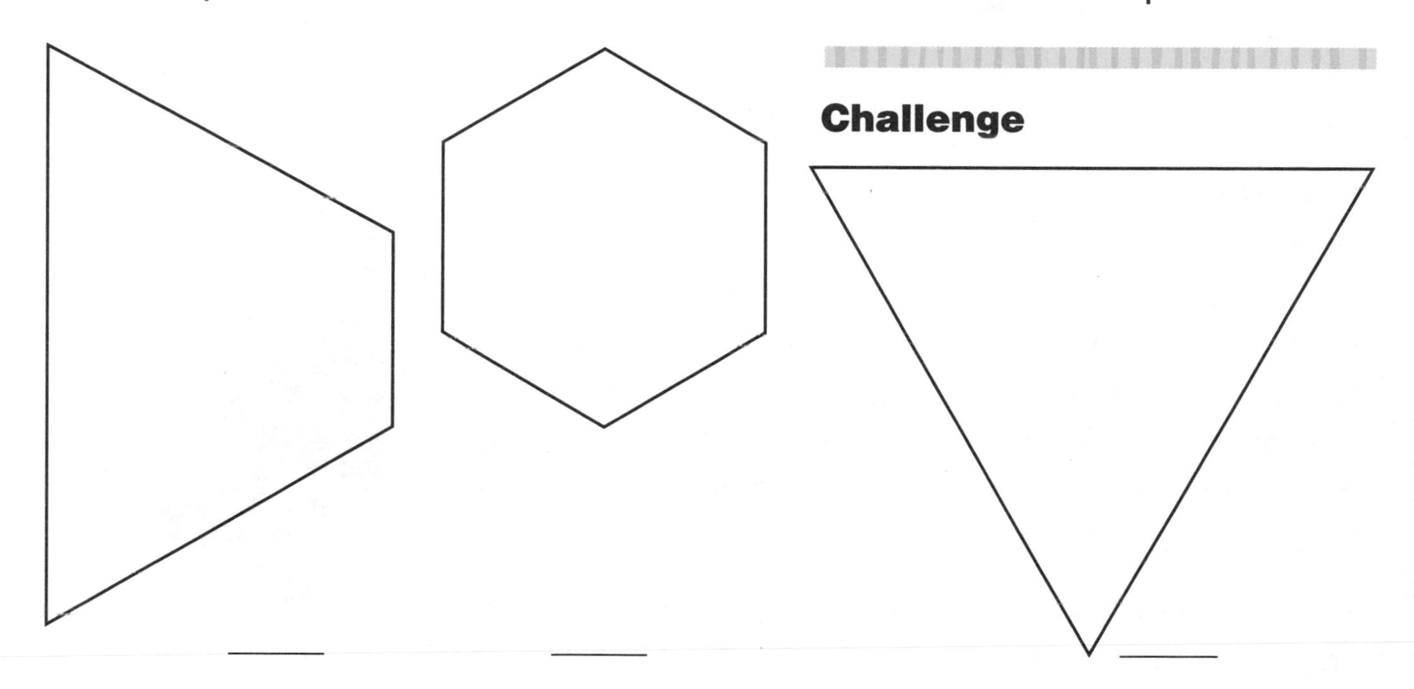

Part 5

Use your Pattern-Block Template to show how you divided the shapes in each section. *Remember:* The number *under* the fraction bar names the number of equal parts into which the whole shape is divided.

Follow-Up

Get together with the rest of the group.

- Compare your answers.
- Use the blocks to check your answers.
- Can more than one fraction be correct?

Date Time

Dressing for the Party

Work in a group of four.

Materials
- ❑ *Math Masters,* p. 129 (Pants and Socks Cutouts)
- ❑ scissors
- ❑ tape
- ❑ blue, red, green, and black crayons or coloring pencils

Problem

Pretend that you have 4 pairs of pants: blue, red, green, and black. You also have 4 pairs of socks: blue, red, green, and black. You have been invited to a party. You need to choose a pair of pants and a pair of socks to wear. Of course, both socks must be the same color. For example, the pants could be blue and both socks black.

How many different combinations of these pants and socks are possible?

Strategy

Use the cutouts on *Math Masters,* page 129, and crayons to help you answer the question.

Decide on a good way for your group to share the following work before you start to answer the question.

- Color the pants in the first row blue.
- Color the pants in the second row red.
- Color the pants in the third row green and those in the fourth row black.
- Color the socks in the same way.
- Cut out each pair of pants and each pair of socks.
- Tape together pairs of pants and pairs of socks to show different outfits. Check that you have only one of each outfit.

Date Time

Dressing for the Party (cont.)

Pretend that you have 4 different colors of pants and 4 different colors of socks.

1. How many different combinations of pants and socks did your group find? ______________________

2. Is this all of the possible combinations? __________

3. How do you know?

____

____

____

____

4. How did your group divide up the work?

____

____

____

____

5. How did your group solve the problem?

____

____

____

____

Date   Time

Math Boxes 8.2

1. Circle $\frac{5}{10}$ of the collection of triangles.

△ △ △ △ △
△ △ △ △ △

Name the fraction that is left in 2 ways.

_____ and _____

SRB 24

2. Number of children per classroom:

25, 30, 26, 28, 33, 35, 28

Median: _____

Maximum: _____

Minimum: _____

Range: _____

SRB 73 74

3. Put in the parentheses needed to complete the number models.

$31 = 3 + 7 \times 4$

$40 = 3 + 7 \times 4$

$4 \times 8 + 2 \times 2 = 36$

$4 \times 8 + 2 \times 2 = 80$

SRB 16 17

4. Subtract.

$$\begin{array}{r} 439 \\ -\ 378 \\ \hline \end{array} \qquad \begin{array}{r} 4,666 \\ -\ 1,297 \\ \hline \end{array} \qquad \begin{array}{r} 3,408 \\ -\ 571 \\ \hline \end{array}$$

SRB 54 55

5. Fill in the missing numbers.

×, ÷	700	60
8		
	4,900	

6. Complete.

24 inches = __________ feet

30 cm = __________ mm

__________ yards = 12 feet

__________ yards = 72 inches

4 meters = __________ centimeters

SRB 122 128 129

Date Time

Fraction Number-Line Poster

1 Whole
Halves
Fourths
Eighths
Thirds
Sixths

Date Time

Math Boxes 8.3

MATH

1. Shade $\frac{3}{8}$ of the circle.

What fraction is *un*shaded? _____

SRB 22 23

2. Fill in the missing numbers.

×, ÷	4	9
	28	
8		72

3. Measure the line segment to the nearest $\frac{1}{4}$ inch.

Draw a line segment $1\frac{3}{4}$ inches long.

SRB 125 126

4. Circle the digit in the millions place.

Put an X on the digit in the ten-thousands place.

Put a box around the digit in the hundreds place.

4 , 9 0 2 , 5 6 7

SRB 18 19

5. Draw an angle that measures between 5° and 90°.

SRB 149–151

6. Write a definition for *parallel.*

____

____

____

____

Write a definition for *intersect.*

____

____

SRB 91

Date Time

Table of Equivalent Fractions

Use your deck of Fraction Cards to find equivalent fractions.
Record them in the table.

Fraction	Equivalent Fractions
$\frac{0}{2}$	
$\frac{1}{2}$	
$\frac{2}{2}$	
$\frac{1}{3}$	
$\frac{2}{3}$	
$\frac{1}{4}$	
$\frac{3}{4}$	
$\frac{1}{5}$	
$\frac{4}{5}$	
$\frac{1}{6}$	
$\frac{5}{6}$	

Challenge

Fraction	Equivalent Fractions
$\frac{1}{8}$	
$\frac{5}{8}$	
$\frac{2}{9}$	

Date Time

Fractions of Sets

What fraction does each picture show? Shade the oval next to each correct answer. There may be more than one correct answer.

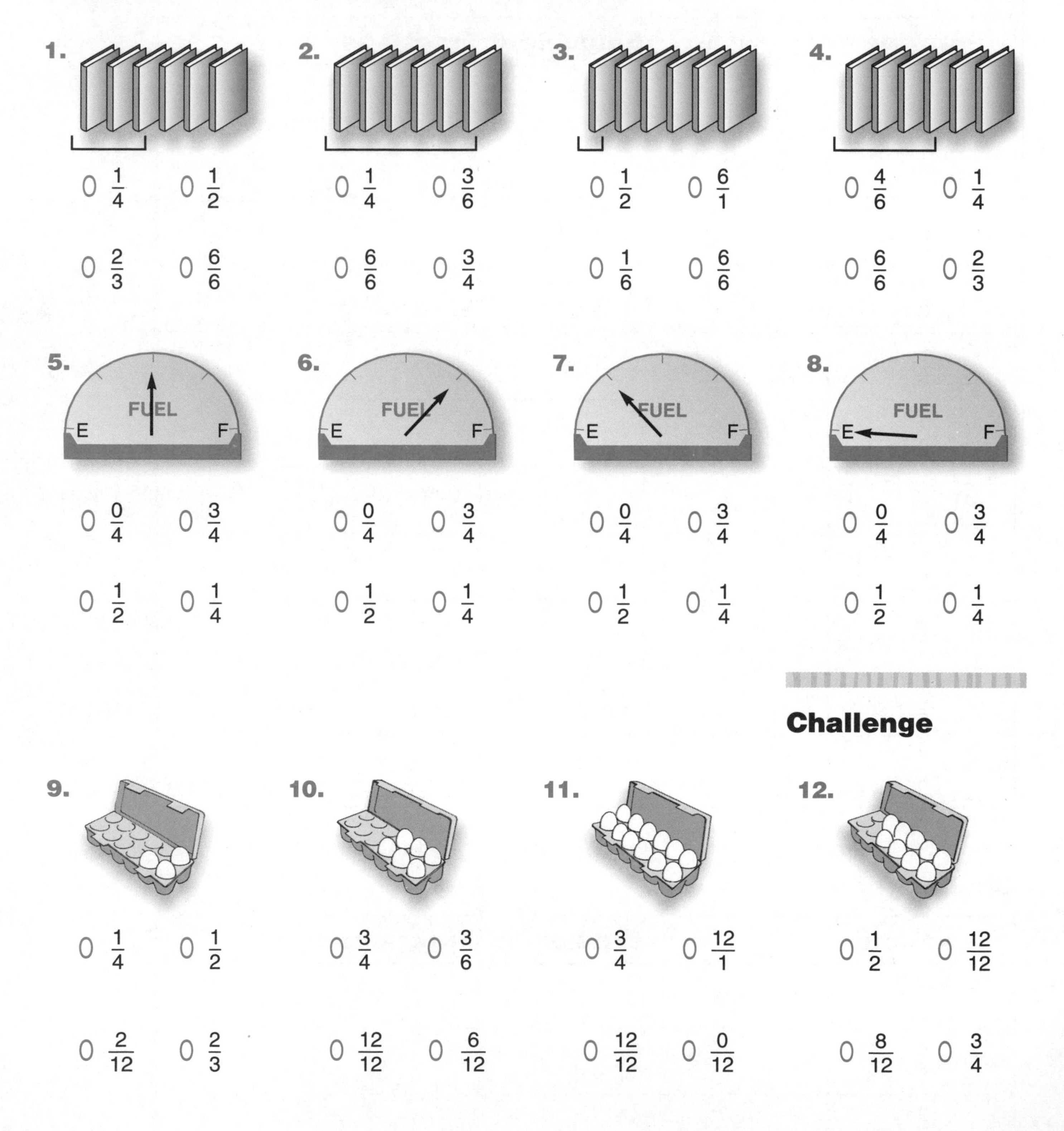

Math Boxes 8.4

MATH

1. Shade $\frac{7}{10}$ of the hats.

SRB 24

2.

$$\begin{array}{r} 3,333 \\ +\ \ 999 \\ \hline \end{array}$$

Write a number model for your estimate.

______ + ______ = ______

Answer: ______

SRB 51 52 168

3. Fill in the missing numbers. Use fractions.

0 ______ $\frac{1}{2}$ ______ 1

4. Write <, >, or =.

0.75 ☐ 0.57

0.09 ☐ 0.9

0.062 ☐ 0.107

12.4 ☐ 14.2

SRB 36

5. Fill in the missing numbers.

×, ÷		600
50	1,500	
		42,000

6. Measure the line segment to the nearest $\frac{1}{4}$ inch.

______

Draw a line segment 1 inch long.

SRB 125 126

Date Time

Color the Fraction Cat

Color the picture below. Follow the color key. For example, all parts with a fraction equivalent to $\frac{2}{3}$ should be colored orange. So the part with $\frac{4}{6}$ should also be colored orange.

Fractions Equal To	Color Key
$\frac{1}{2}$	yellow
$\frac{2}{3}$	orange
$\frac{1}{3}$	brown
$\frac{1}{4}$	green
$\frac{3}{4}$	black

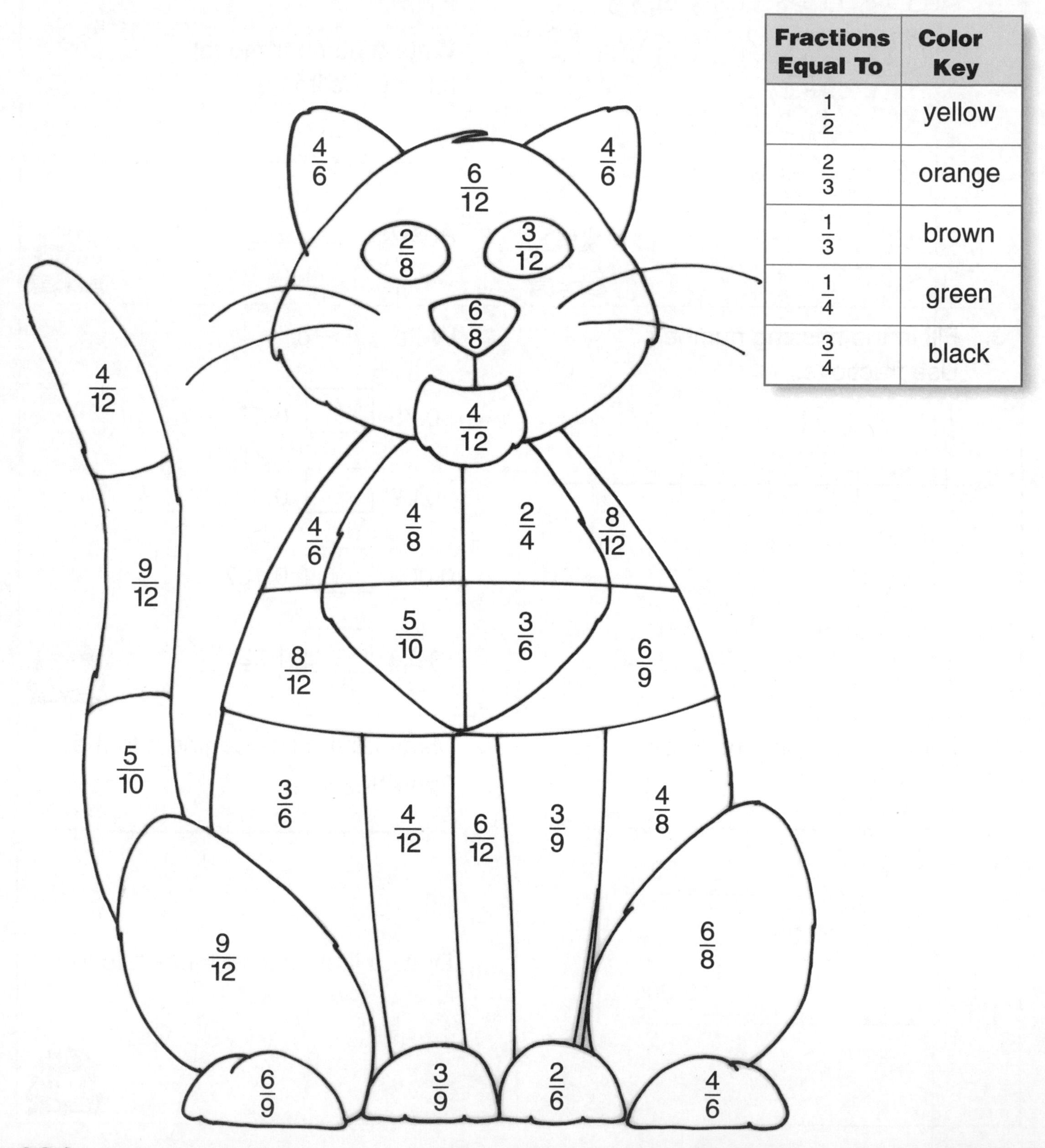

 Use with Lesson 8.5.

Date Time

Math Boxes 8.5

1. Color $\frac{2}{5}$ of the rectangle.

What fraction is *un*colored? _____

SRB 22 23

2. Solve.

Unit: stamps

$54 \div 9 =$ _____

$27 \div 3 =$ _____

_____ $= 36 \div 6$

_____ $= 64 \div 8$

$45 \div 5 =$ _____

SRB 46 47

3. Fill in the missing numbers. Use fractions.

0, _____, $\frac{2}{3}$, _____, _____

SRB

4. Complete the number models.

Unit: books

$(4 + 3) - 2 =$ _____

$10 = 6 + (2 +$ _____$)$

_____ $= 3 \times (9 - 0)$

$(5 \times 5) - 4 =$ _____

SRB 16 17

5. Write 4 fractions equivalent to $\frac{1}{2}$.

_____ _____

_____ _____

SRB 30

6. Subtract.

$$\begin{array}{r} 6{,}000 \\ -\ \ 583 \\ \hline \end{array} \qquad \begin{array}{r} 801 \\ -\ 472 \\ \hline \end{array} \qquad \begin{array}{r} 3{,}411 \\ -\ 2{,}862 \\ \hline \end{array}$$

SRB 54 55

Date Time

More Than ONE

Use the circles that you cut out for the Math Message.

1. Glue 3 halves into the two whole circles.

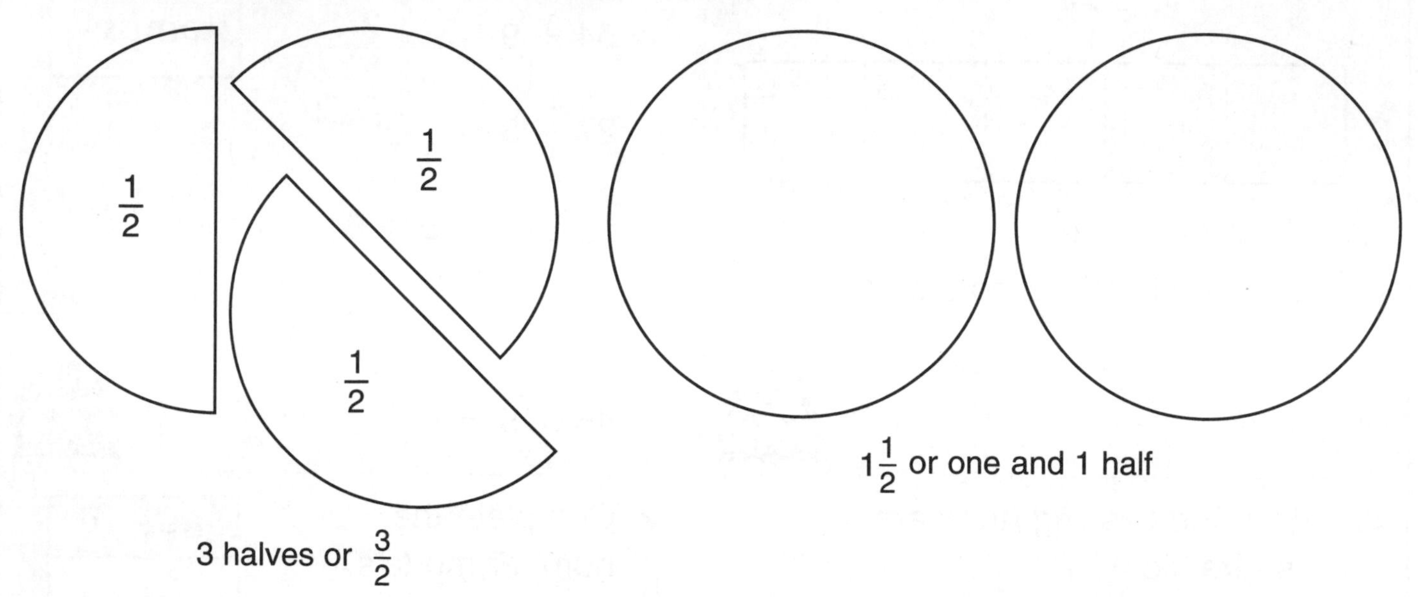

2. Glue 6 fourths into the two whole circles. Fill in the missing digits in the question, the fraction, and the mixed number.

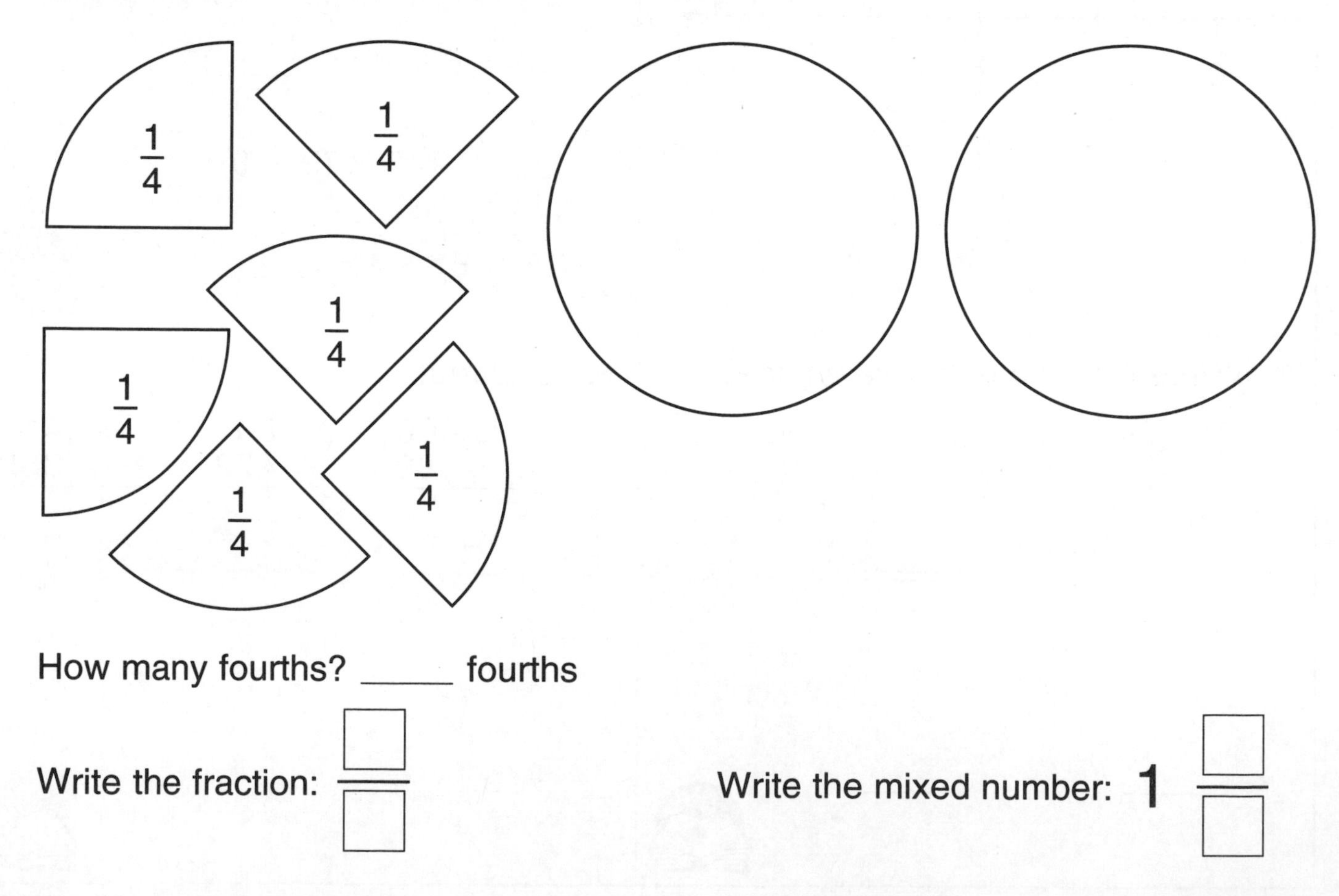

How many fourths? _____ fourths

Write the fraction: $\frac{\square}{\square}$

Write the mixed number: $1\frac{\square}{\square}$

Date Time

More Than ONE (cont.)

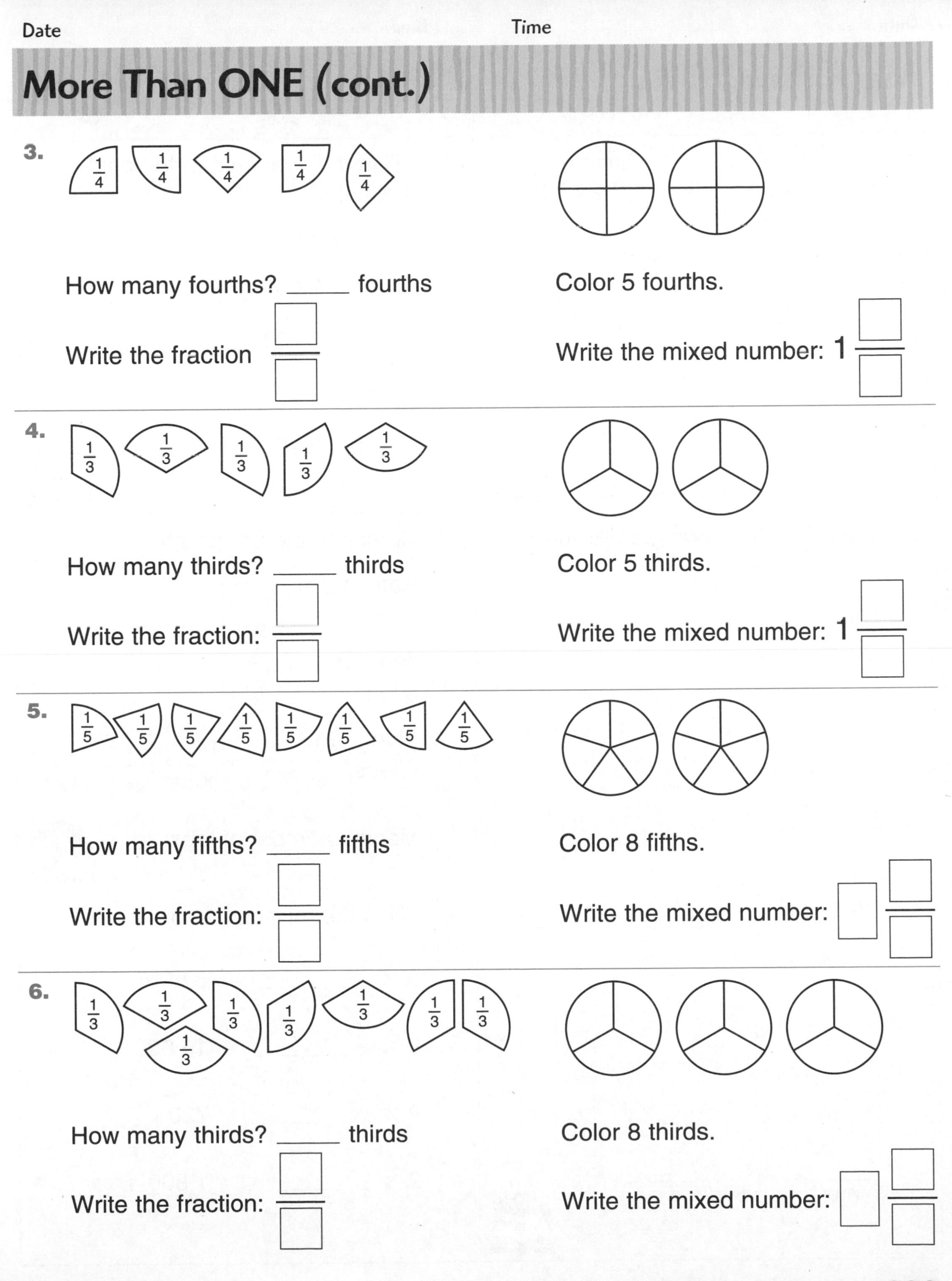

3. $\frac{1}{4}$ $\frac{1}{4}$ $\frac{1}{4}$ $\frac{1}{4}$ $\frac{1}{4}$

How many fourths? _____ fourths

Write the fraction $\frac{\square}{\square}$

Color 5 fourths.

Write the mixed number: $1\frac{\square}{\square}$

4. $\frac{1}{3}$ $\frac{1}{3}$ $\frac{1}{3}$ $\frac{1}{3}$ $\frac{1}{3}$

How many thirds? _____ thirds

Write the fraction: $\frac{\square}{\square}$

Color 5 thirds.

Write the mixed number: $1\frac{\square}{\square}$

5. $\frac{1}{5}$ $\frac{1}{5}$ $\frac{1}{5}$ $\frac{1}{5}$ $\frac{1}{5}$ $\frac{1}{5}$ $\frac{1}{5}$ $\frac{1}{5}$

How many fifths? _____ fifths

Write the fraction: $\frac{\square}{\square}$

Color 8 fifths.

Write the mixed number: $\square\frac{\square}{\square}$

6. $\frac{1}{3}$ $\frac{1}{3}$ $\frac{1}{3}$ $\frac{1}{3}$ $\frac{1}{3}$ $\frac{1}{3}$ $\frac{1}{3}$ $\frac{1}{3}$

How many thirds? _____ thirds

Write the fraction: $\frac{\square}{\square}$

Color 8 thirds.

Write the mixed number: $\square\frac{\square}{\square}$

Date Time

Math Boxes 8.6

1. Draw a set of 12 Xs. Circle 9 of them. What fraction of the whole set are the 9 Xs?

______

SRB 24

2. Fill in the oval next to the best estimate.

589 + 2,115 = ______

0 about 2,000

0 about 2,500

0 about 2,200

0 about 2,700

SRB 168

3. Write four fractions greater than $\frac{1}{3}$.

____ ____

____ ____

4. Complete the bar graph.

Kate swam 5 laps.

Jen swam 3 laps.

Marc swam 6 laps.

Laps Swum	Kate	Jen	Marc
6			
5			
4			
3			
2			
1			
0			

Median number of laps: ____

SRB 78 80 81

5. Write 4 fractions equivalent to $\frac{1}{4}$.

____ ____

____ ____

SRB 30

6. Fill in the missing factor.

$6 \times$ ______ $= 3,600$

$8 \times$ ______ $= 16,000$

$9 \times$ ______ $= 720$

$2 \times$ ______ $= 1,800$

Date Time

Fraction Number Stories

Solve these number stories. Use pennies or counters, or draw pictures to help you.

1. There are 8 apples in the package. Glenn did not eat any. What fraction of the package did Glenn eat?

   ____

2. Anik bought a dozen eggs at the supermarket. When he got home, he found that $\frac{1}{6}$ of the eggs were cracked. How many eggs were cracked?

   ____ eggs

3. Chante used $\frac{2}{3}$ of a package of ribbon to wrap presents. Did she use more or less than $\frac{3}{4}$ of the package?

   ____

4. I had 2 whole cookies. I gave you $\frac{1}{4}$ of 1 cookie. How many cookies did I have left?

   ____ cookies

5. There are 10 quarters. You have 3. I have 2. What fraction of the quarters do you have?

   ____

   What fraction of the quarters do I have?

   ____

   What fraction of the quarters do we have together?

   ____

6. One day, Edwin read $\frac{1}{3}$ of a book. The next day, he read another $\frac{1}{3}$ of the book. What fraction of the book had he read after 2 days?

   ____

   What fraction of the book did he have left to read?

   ____

7. Dorothy walks $1\frac{1}{2}$ miles to school. Jaime walks $1\frac{2}{4}$ miles to school. Who walks the longer distance?

   ____

   ____

8. Twelve children shared 2 medium-size pizzas equally. What fraction of 1 whole pizza did each child eat?

   ____

Date Time

Fraction Number Stories (cont.)

9. Write a fraction story. Ask your partner to solve it.

____

____

____

____

____

Draw eggs in each carton to show the fraction.

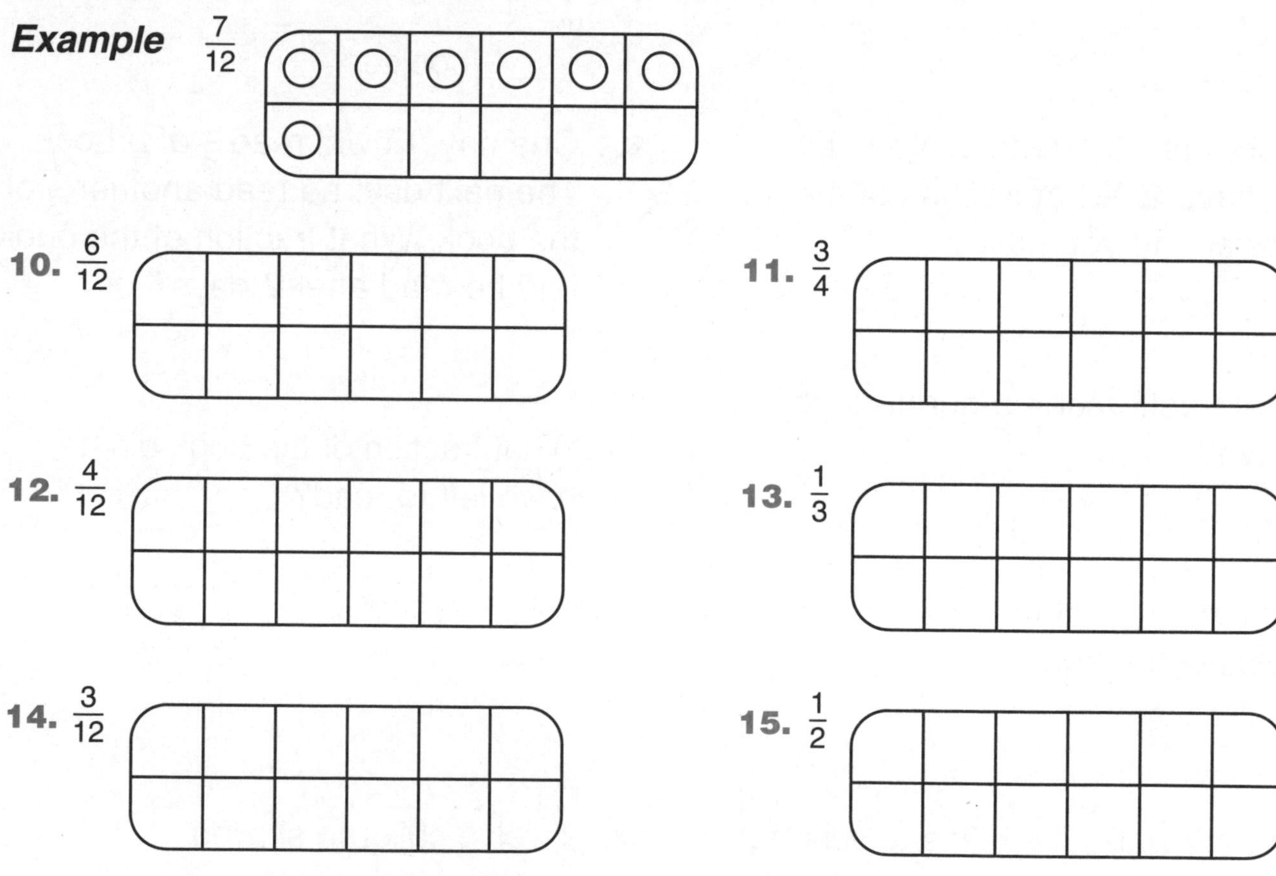

16. Julie drank $\frac{1}{4}$ of a glass of juice.

Draw an empty glass.

Shade in the glass to show how much juice is left.

Date Time

Math Boxes 8.7

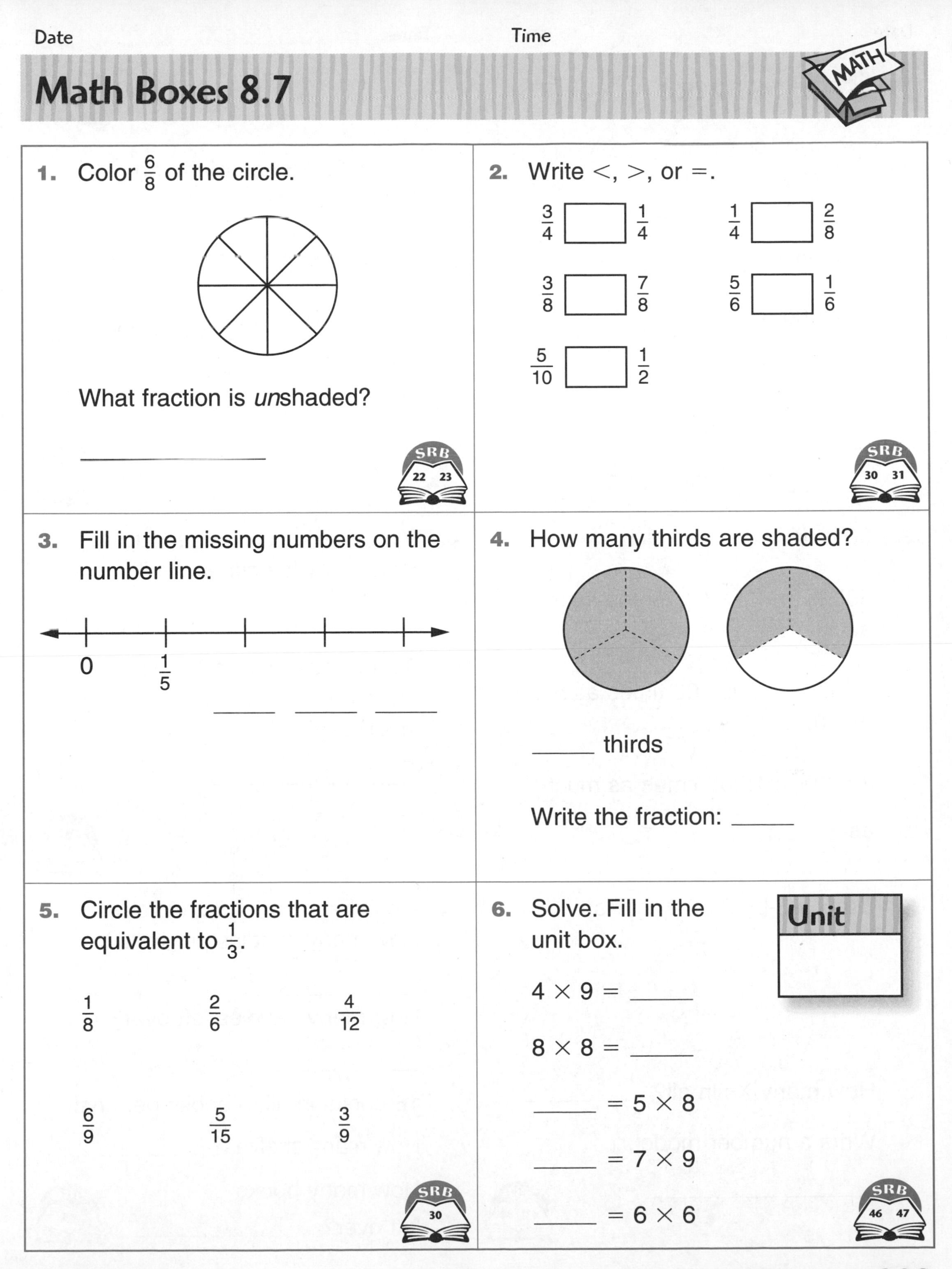

1. Color $\frac{6}{8}$ of the circle.

What fraction is *un*shaded?

______

SRB 22 23

2. Write $<$, $>$, or $=$.

$\frac{3}{4}$ ☐ $\frac{1}{4}$

$\frac{1}{4}$ ☐ $\frac{2}{8}$

$\frac{3}{8}$ ☐ $\frac{7}{8}$

$\frac{5}{6}$ ☐ $\frac{1}{6}$

$\frac{5}{10}$ ☐ $\frac{1}{2}$

SRB 30 31

3. Fill in the missing numbers on the number line.

0, $\frac{1}{5}$, ____ ____ ____

4. How many thirds are shaded?

____ thirds

Write the fraction: ____

5. Circle the fractions that are equivalent to $\frac{1}{3}$.

$\frac{1}{8}$ $\frac{2}{6}$ $\frac{4}{12}$

$\frac{6}{9}$ $\frac{5}{15}$ $\frac{3}{9}$

SRB 30

6. Solve. Fill in the unit box.

Unit

$4 \times 9 =$ ____

$8 \times 8 =$ ____

____ $= 5 \times 8$

____ $= 7 \times 9$

____ $= 6 \times 6$

SRB 46 47

Date Time

Math Boxes 8.8

1. Solve.

$5 \times 9 =$ ________

$5 \times 90 =$ ________

$5 \times 900 =$ ________

________ $= 3 \times 8$

________ $= 30 \times 80$

________ $= 300 \times 80$

2. Share \$3.75 equally among 3 people.

Each person gets \$________ .

Share \$10.00 equally among 4 people.

Each person gets \$________ .

SRB 67

3. 30 is 10 times as much as ______.

500 is ________ times as much as 5.

________ is 100 times as much as 80.

40,000 is 1,000 times as much as ________.

4. 9 cups. 9 ice cubes per cup. How many ice cubes in all?

____

3 packages. 9 juice boxes per package. How many juice boxes in all?

____

SRB 65

5. Draw a 3-by-8 array of Xs.

How many Xs in all? ________

Write a number model.

____

SRB 63 64

6. 9 children share 18 candies. How many candies per child?

____

How many candies left over?

____

16 books in all. 3 books per shelf.

How many shelves? ________

How many books left over? ________

SRB 68

Multiples of 10, 100, and 1,000

Solve each problem.

1. a. 7 [40s] = ________ b. $7 \times 40 =$ ________

2. a. 600 [20s] – ________ b. $600 \times 20 =$ ________

3. a. How many 3s are in 2,700? ________

   b. ________ $\times 3 = 2{,}700$ c. $2{,}700 \div 3 =$ ________

4. How many 50s are in 4,000? ______

5. How many 800s are in 2,400? ______

6. How many 70s are in 420? ______

7. a. $40 \times 300 =$ ________ b. $12{,}000 \div 40 =$ ________

For Problems 8–11, use the information on the next two journal pages.

8. a. Which animal might weigh about 20 times as much as a 30-pound raccoon? ________________

   b. Can you name two other animals which might weigh 20 times as much as a 30-pound raccoon?

   ____

9. About how many 200-pound American alligators weigh about as much as a 3,200-pound Beluga whale? ________________

Challenge

10. Which animal might weigh about 100 times as much as the combined weights of a 15-pound Arctic fox and a 10-pound Arctic fox? ________________

11. About how many $4\frac{1}{2}$-pound snowshoe hares weigh as much as a 27-pound porcupine? ________________

Date Time

Adult Weights of North American Animals

Source: *Do Elephants Eat Too Much?* Robert Balfanz. Everyday Learning Corporation, 1992.

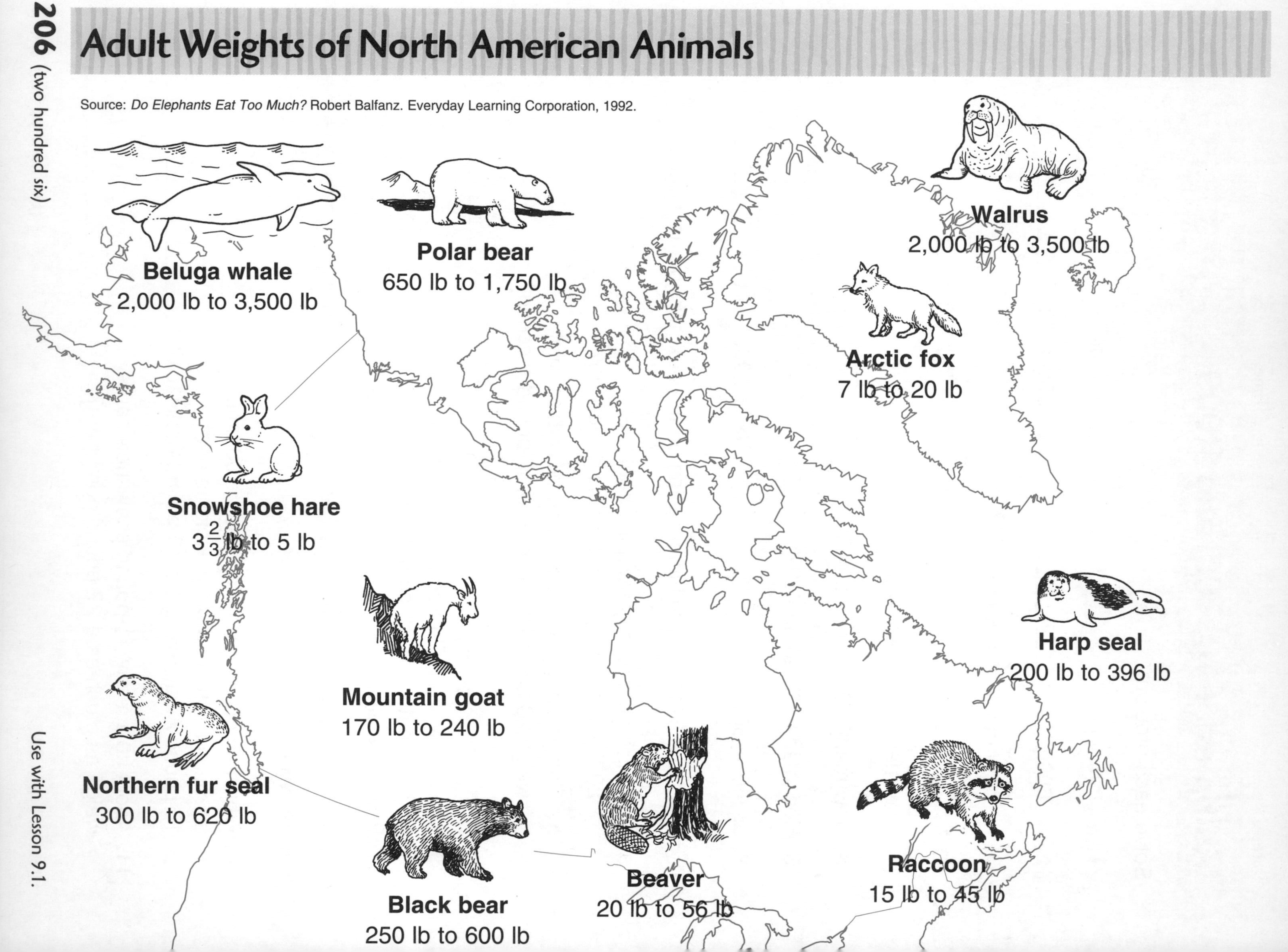

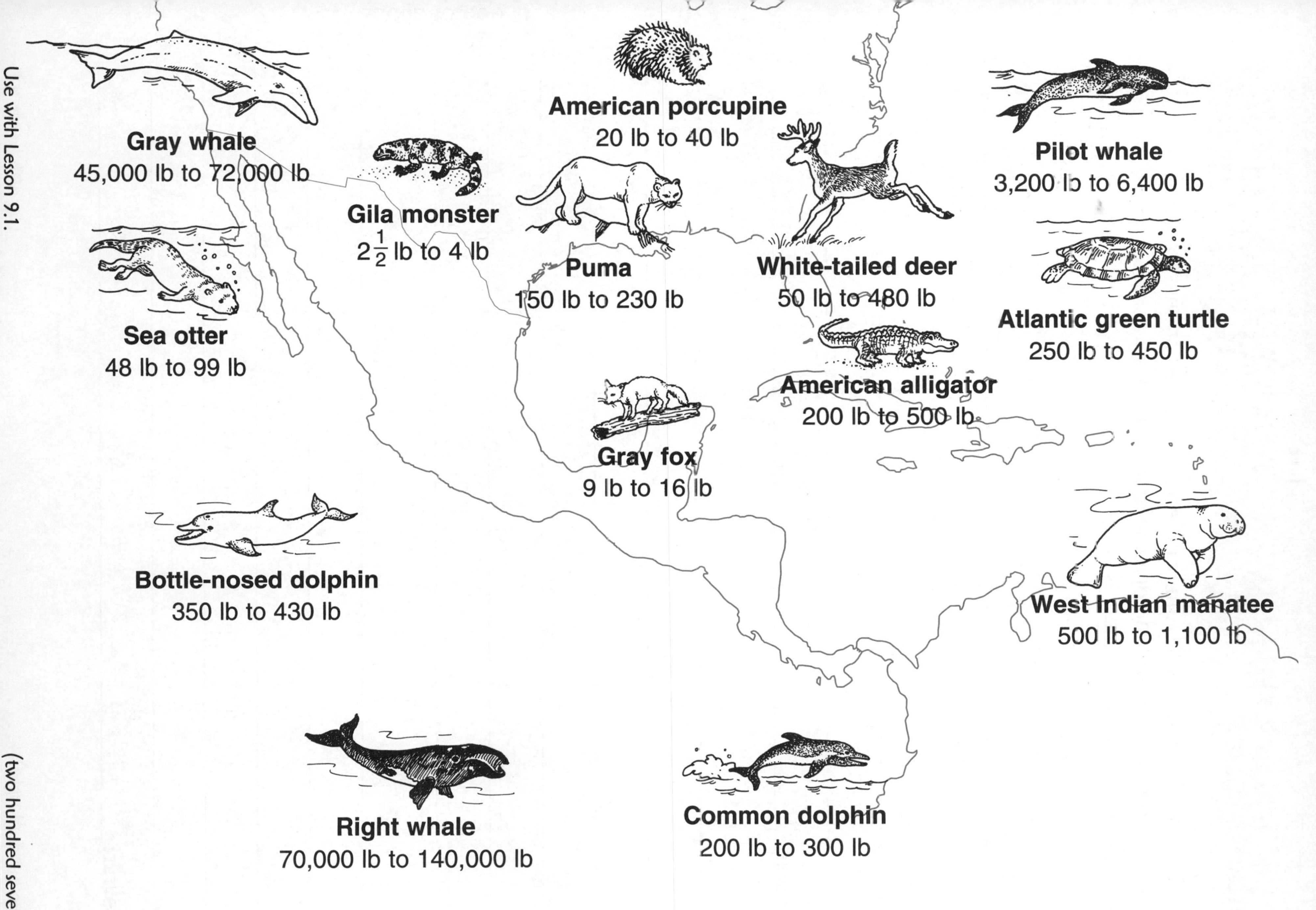

Gray whale
45,000 lb to 72,000 lb
Gila monster
$2\frac{1}{2}$ lb to 4 lb
American porcupine
20 lb to 40 lb
Puma
150 lb to 230 lb
White-tailed deer
50 lb to 480 lb
Pilot whale
3,200 lb to 6,400 lb
Sea otter
48 lb to 99 lb
Atlantic green turtle
250 lb to 450 lb
American alligator
200 lb to 500 lb
Gray fox
9 lb to 16 lb
Bottle-nosed dolphin
350 lb to 430 lb
West Indian manatee
500 lb to 1,100 lb
Right whale
70,000 lb to 140,000 lb
Common dolphin
200 lb to 300 lb

Date Time

Number Stories

Use the Adult Weights of North American Animals poster on the previous pages. Make up multiplication and division animal number stories. Ask a partner to solve your number stories.

1. ______________________________

____

____

____

____

Answer: ______________________________

2. ______________________________

____

____

____

____

Answer: ______________________________

3. ______________________________

____

____

____

____

Answer: ______________________________

Date Time

Math Boxes 9.1

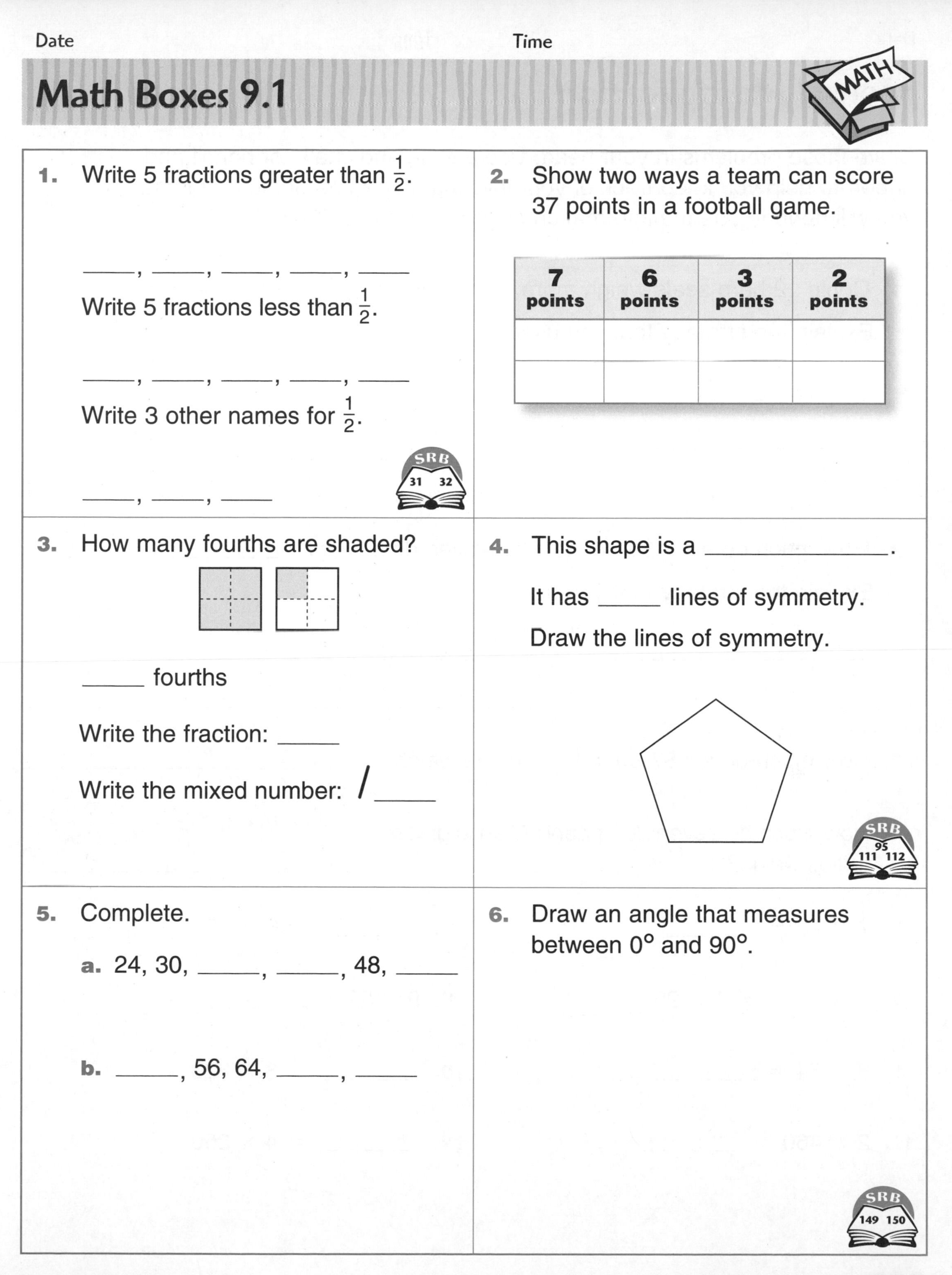

1. Write 5 fractions greater than $\frac{1}{2}$.

_____, _____, _____, _____, _____

Write 5 fractions less than $\frac{1}{2}$.

_____, _____, _____, _____, _____

Write 3 other names for $\frac{1}{2}$.

_____, _____, _____

SRB 31 32

2. Show two ways a team can score 37 points in a football game.

7 points	6 points	3 points	2 points

3. How many fourths are shaded?

_____ fourths

Write the fraction: _____

Write the mixed number: _____

4. This shape is a _______________.

It has _____ lines of symmetry.

Draw the lines of symmetry.

SRB 95 111 112

5. Complete.

a. 24, 30, _____, _____, 48, _____

b. _____, 56, 64, _____, _____

6. Draw an angle that measures between 0° and 90°.

SRB 149 150

Date Time

Mental Multiplication

Solve these problems in your head. Use a slate and chalk, or pencil and paper, to help you keep track of your thinking. For some of the problems, you will need to use the information on journal pages 206 and 207.

1. Could 12 harp seals weigh more than 1 ton? ______ Less than 1 ton? ______

   Explain the strategy that you used.

   ____

   ____

   ____

2. How much do eight 53-pound white-tailed deer weigh? ______________

   Explain the strategy that you used.

   ____

   ____

3. How much do six 87-pound sea otters weigh? ______________

4. How much do seven 260-pound Atlantic green turtles weigh? ______________

5. $7 \times 23 =$ ________

6. ________ $= 8 \times 46$

7. ________ $= 4 \times 26$

8. $9 \times 32 =$ ________

9. $6 \times 54 =$ ________

10. ________ $= 3 \times 320$

11. $2 \times 460 =$ ________

12. ________ $= 4 \times 250$

Date Time

Which Is the Best Buy?

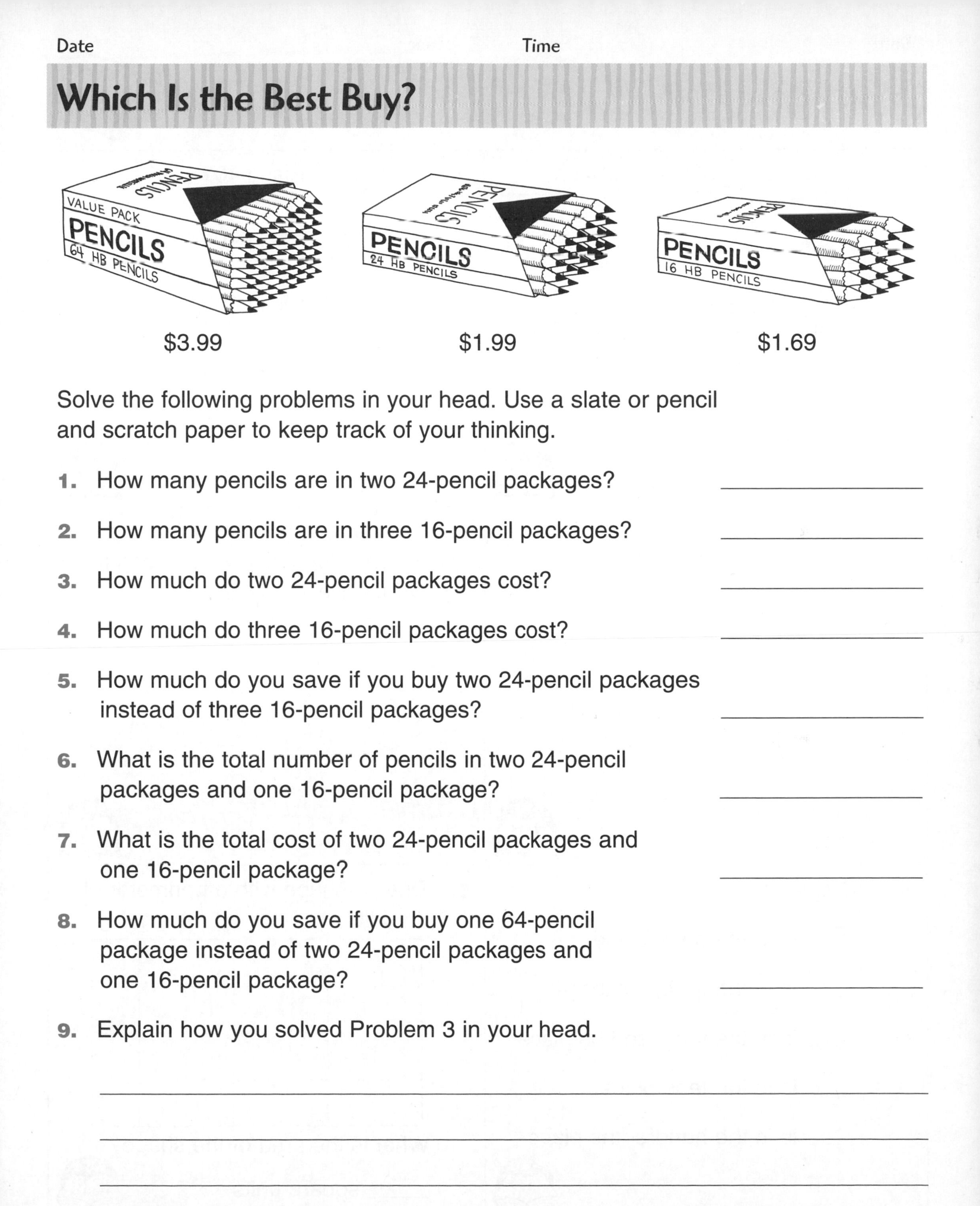

$3.99 $1.99 $1.69

Solve the following problems in your head. Use a slate or pencil and scratch paper to keep track of your thinking.

1. How many pencils are in two 24-pencil packages? ______________
2. How many pencils are in three 16-pencil packages? ______________
3. How much do two 24-pencil packages cost? ______________
4. How much do three 16-pencil packages cost? ______________
5. How much do you save if you buy two 24-pencil packages instead of three 16-pencil packages? ______________
6. What is the total number of pencils in two 24-pencil packages and one 16-pencil package? ______________
7. What is the total cost of two 24-pencil packages and one 16-pencil package? ______________
8. How much do you save if you buy one 64-pencil package instead of two 24-pencil packages and one 16-pencil package? ______________
9. Explain how you solved Problem 3 in your head.

____

____

____

Date Time

Math Boxes 9.2

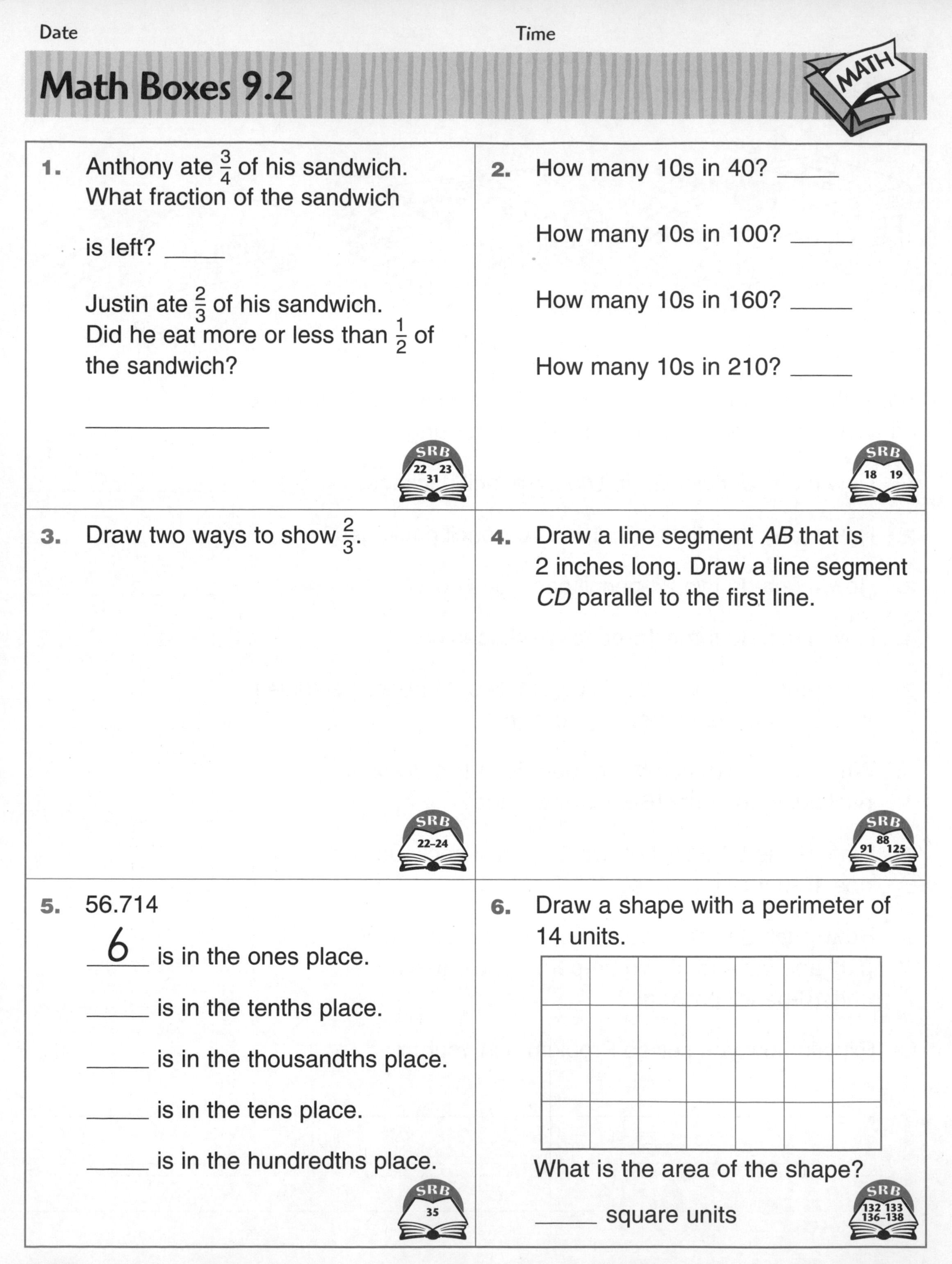

1. Anthony ate $\frac{3}{4}$ of his sandwich. What fraction of the sandwich is left? _____

Justin ate $\frac{2}{3}$ of his sandwich. Did he eat more or less than $\frac{1}{2}$ of the sandwich?

____

SRB 22 23 31

2. How many 10s in 40? _____

How many 10s in 100? _____

How many 10s in 160? _____

How many 10s in 210? _____

SRB 18 19

3. Draw two ways to show $\frac{2}{3}$.

SRB 22–24

4. Draw a line segment *AB* that is 2 inches long. Draw a line segment *CD* parallel to the first line.

SRB 88 91 125

5. 56.714

__6__ is in the ones place.

_____ is in the tenths place.

_____ is in the thousandths place.

_____ is in the tens place.

_____ is in the hundredths place.

SRB 35

6. Draw a shape with a perimeter of 14 units.

What is the area of the shape?

_____ square units

SRB 132 133 136–138

Date Time

Array Multiplication 1

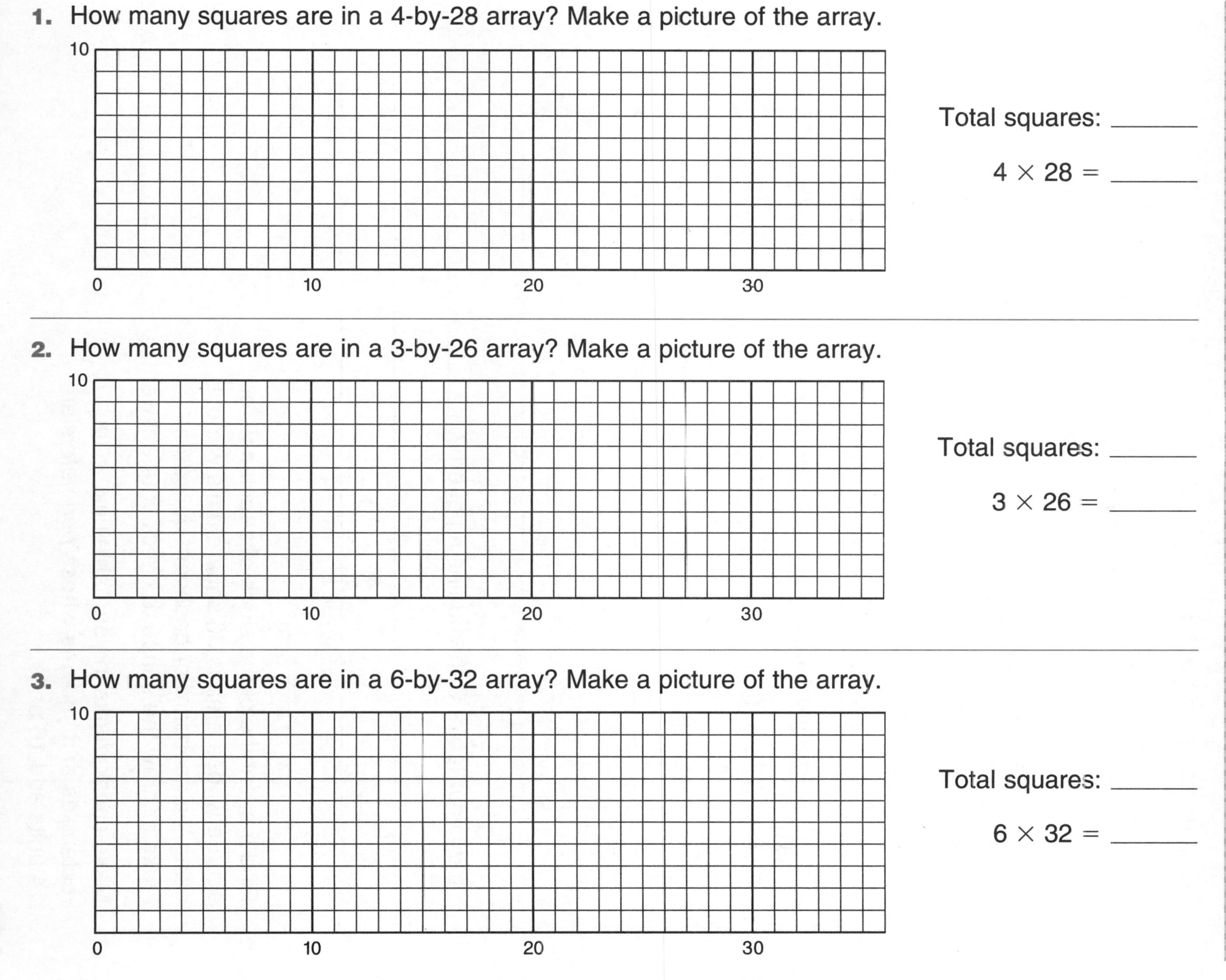

1. How many squares are in a 4-by-28 array? Make a picture of the array.

Total squares: ______

$4 \times 28 =$ ______

2. How many squares are in a 3-by-26 array? Make a picture of the array.

Total squares: ______

$3 \times 26 =$ ______

3. How many squares are in a 6-by-32 array? Make a picture of the array.

Total squares: ______

$6 \times 32 =$ ______

Geoboard Areas

Record your results in this table.

Geoboard Areas		
Area	Longer Sides	Shorter Sides
12 square units	units	units
12 square units	units	units
6 square units	units	units
6 square units	units	unit
16 square units	units	units
16 square units	units	units

1. Study your table. Can you find a pattern? ______________________

____

____

____

2. Find the lengths of the sides of a rectangle or square whose area is 30 square units without using the geoboard or geoboard dot paper. Make or draw the shape to check your answer. ______________________

3. Make check marks in your table next to the rectangles and squares whose *perimeters* are 14 units and 16 units.

Date Time

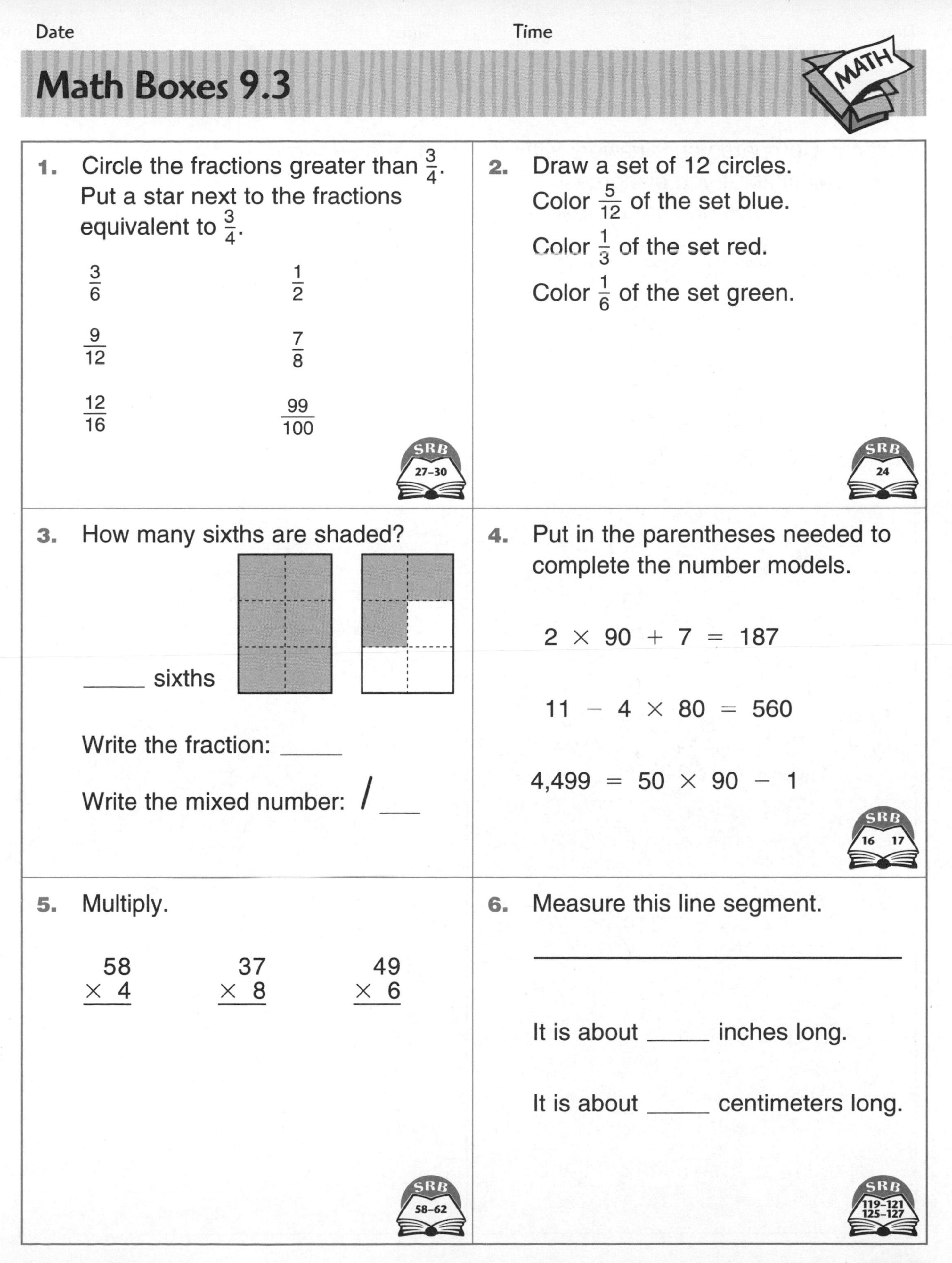

Math Boxes 9.3

MATH

1. Circle the fractions greater than $\frac{3}{4}$. Put a star next to the fractions equivalent to $\frac{3}{4}$.

$\frac{3}{6}$ $\frac{1}{2}$

$\frac{9}{12}$ $\frac{7}{8}$

$\frac{12}{16}$ $\frac{99}{100}$

SRB 27–30

2. Draw a set of 12 circles.

Color $\frac{5}{12}$ of the set blue.

Color $\frac{1}{3}$ of the set red.

Color $\frac{1}{6}$ of the set green.

SRB 24

3. How many sixths are shaded?

_____ sixths

Write the fraction: _____

Write the mixed number: / ___

4. Put in the parentheses needed to complete the number models.

$2 \times 90 + 7 = 187$

$11 - 4 \times 80 = 560$

$4{,}499 = 50 \times 90 - 1$

SRB 16 17

5. Multiply.

$$\begin{array}{r} 58 \\ \times\ 4 \\ \hline \end{array} \qquad \begin{array}{r} 37 \\ \times\ 8 \\ \hline \end{array} \qquad \begin{array}{r} 49 \\ \times\ 6 \\ \hline \end{array}$$

SRB 58–62

6. Measure this line segment.

It is about _____ inches long.

It is about _____ centimeters long.

SRB 119–121 125–127

Date Time

Using the Partial-Products Algorithm

Multiply. Compare your answers with your partner's answers.
Use a calculator if you disagree.
If you made a mistake on a problem, work it again.

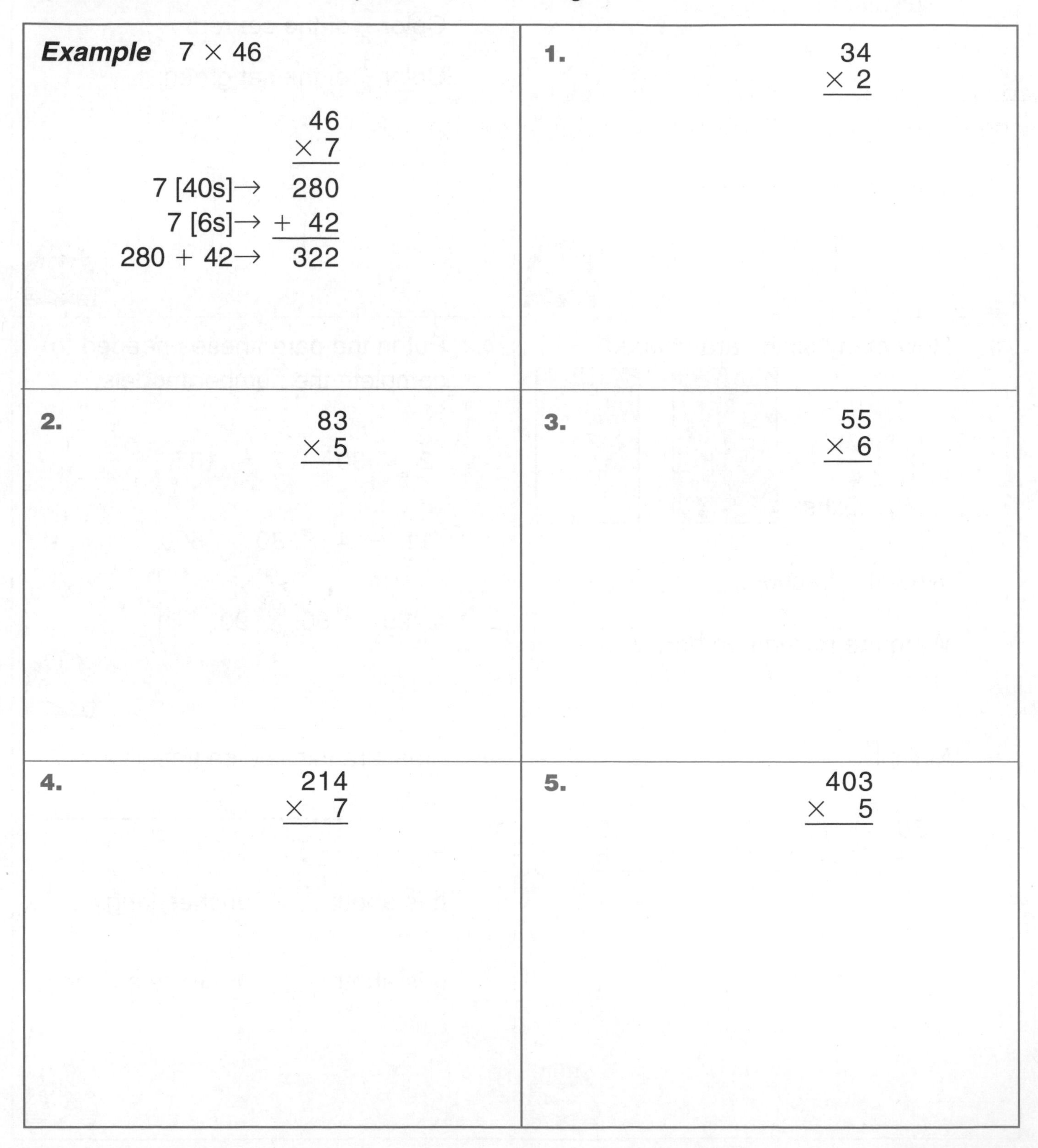

Date Time

Who Am I?

In each riddle, I am a different whole number.
Use the clues to find out who I am.

1. **Clue 1:** I am greater than 30 and less than 40.
   Clue 2: The sum of my digits is less than 5.

   Who am I? ____________

2. **Clue 1:** I am greater than 15 and less than 40.
   Clue 2: If you double me, I become a number that ends in 0.
   Clue 3: $\frac{1}{5}$ of me is equal to 5.

   Who am I? ____________

3. **Clue 1:** I am less than 100.
   Clue 2: The sum of my digits is 4.
   Clue 3: Half of me is an odd number.

   Who am I? ____________

4. **Clue 1:** If you multiply me by 2, I become a number greater than 20 and less than 40.
   Clue 2: If you multiply me by 6, I end in 8.
   Clue 3: If you multiply me by 4, I end in 2.

   Who am I? ____________

Challenge

5. **Clue 1:** Double my tens digit to get my ones digit.
   Clue 2: Double me and I am less than 50.

   Who am I? ____________

6. **Clue 1:** Double me, and I am greater than 80 and less than 100.
   Clue 2: If you double me, I end in 4.
   Clue 3: My ones digit is greater than my tens digit.

   Who am I? ____________

Date Time

Math Boxes 9.4

1. There are _____ books in $\frac{2}{5}$ of a set of 25 books.

There are _____ minutes in $\frac{3}{4}$ of an hour.

I have 6 books. This is $\frac{1}{6}$ of a set of books. How many books are in the complete set?

_____ books

2.

	Think: How many...
$63 \div 7 =$ _____	7s in 63?
$630 \div 7 =$ _____	7s in 630?
$48 \div 6 =$ _____	6s in 48?
$480 \div 6 =$ _____	6s in 480?

SRB 46 47

3. The length of the longer side is _____ units.

The length of the shorter side is _____ units.

The area of the rectangle is _____ square units.

SRB 138

4. Complete the "What's My Rule?" table.

in
Rule
Add 25 minutes
out

in	out
7:00	
3:15	
5:45	
	7:40
	11:10

SRB 179 180

5. Suppose you like pizza and are very hungry. Would you rather have $\frac{4}{5}$ of a pizza or $\frac{8}{10}$ of a pizza?

____

Why? _______________

____

____

SRB 27 28

6. Draw an angle that measures between 180° and 270°.

SRB 149 150

Date Time

Shopping at the Stock-Up Sale

Use the Stock-Up Sale Poster #2 on page 241 in the *Student Reference Book.* Solve each number story below. Show how you got the answers.

1. When Mason sees bars of soap at the Stock-Up Sale, he wants to buy at least 5. He has $4.00. If there is no tax, can he buy 5 bars of soap? ________

   Number model: ______________________________

   Can he buy 6 bars? ________

2. Vic's mom gave him a $5.00 bill to buy a toothbrush. If he goes to the sale, can he buy 5 toothbrushes? ________

   Suppose there is no tax. Exactly how much money does Vic need in order to be able to buy 5 toothbrushes at the sale price? ________

   Number model: ______________________________

3. Andrea wants 2 audio tapes. How much more will it cost her to buy 5 tapes at the sale price rather than 2 tapes at the regular price? ____________

4. If the store charges 10 percent sales tax, what will the total cost of the 5 audio tapes be? ________

   Number model: ______________________________

5. Make up a Stock-Up Sale story of your own.

   ____

   ____

   ____

   Answer: ____________

   Number model: ______________________________

Date Time

Using the Partial-Products Algorithm

Multiply. Compare your answers with your partner's answers.
Use a calculator if you disagree.
If you did a problem wrong, work it again.

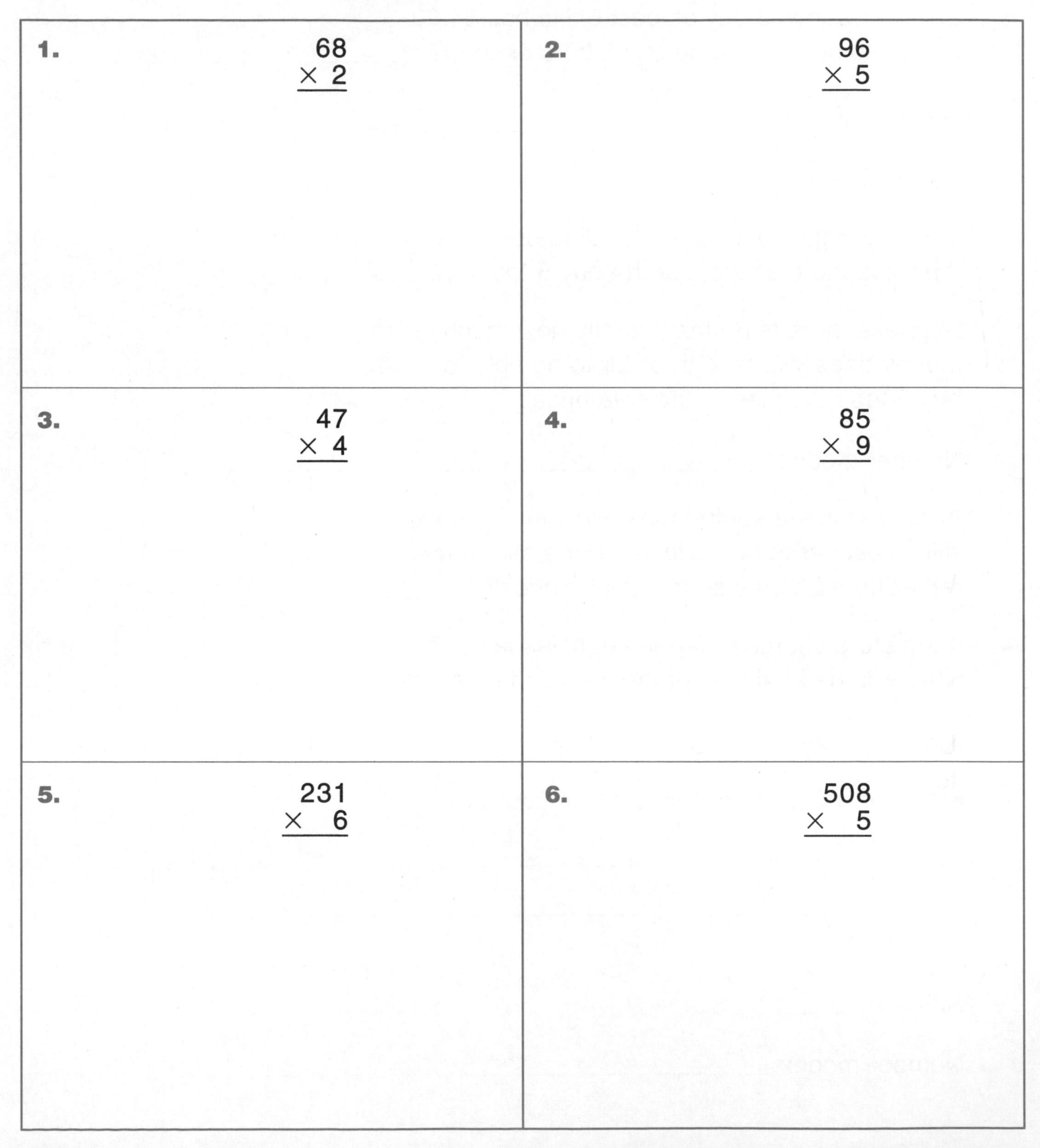

1. $\begin{array}{r} 68 \\ \times\ 2 \\ \hline \end{array}$	**2.** $\begin{array}{r} 96 \\ \times\ 5 \\ \hline \end{array}$
3. $\begin{array}{r} 47 \\ \times\ 4 \\ \hline \end{array}$	**4.** $\begin{array}{r} 85 \\ \times\ 9 \\ \hline \end{array}$
5. $\begin{array}{r} 231 \\ \times\ \ 6 \\ \hline \end{array}$	**6.** $\begin{array}{r} 508 \\ \times\ \ 5 \\ \hline \end{array}$

Date Time

Math Boxes 9.5

MATH

1. Estimate the cost of these items:

4 giant stickers at $0.88 each

about $_____._____

2 packs of file cards at $1.69 each

about $_____._____

SRB 167

2. Fill in the unit box. Solve.

Unit

$49 \div 7 -$ _____

$36 \div 9 =$ _____

$54 \div 6 =$ _____

_____ $= 40 \div 8$

_____ $\div 8 = 8$

SRB 46 47

3. What 3-D shape is this a picture of?

0 sphere

0 cylinder

0 pyramid

What is the shape of the base?

____

SRB 100 101 107

4. Solve.

$$\begin{array}{r} 678 \\ +\ 492 \\ \hline \end{array} \qquad \begin{array}{r} 704 \\ -\ 358 \\ \hline \end{array}$$

SRB 51–57

5. Use the partial-products algorithm to solve.

$$\begin{array}{r} 49 \\ \times\ 7 \\ \hline \end{array} \qquad \begin{array}{r} 652 \\ \times\ \ 3 \\ \hline \end{array} \qquad \begin{array}{r} 408 \\ \times\ \ 8 \\ \hline \end{array}$$

SRB 58 59

6. Fill in the empty frames and the rule box.

Rules: + 40 | _____

Frames: _____ → 83 → _____ → 98 → 73 → _____

SRB 176 177

Date Time

Factor Bingo Game Mat

Write any of the numbers 2–90 onto the grid above.

You may use a number only once.

To help you keep track of the numbers you use, circle them in the list.

	2	3	4	5	6	7	8	9	10
11	12	13	14	15	16	17	18	19	20
21	22	23	24	25	26	27	28	29	30
31	32	33	34	35	36	37	38	39	40
41	42	43	44	45	46	47	48	49	50
51	52	53	54	55	56	57	58	59	60
61	62	63	64	65	66	67	68	69	70
71	72	73	74	75	76	77	78	79	80
81	82	83	84	85	86	87	88	89	90

Date Time

Using the Partial-Products Algorithm

Multiply. Compare your answers with your partner's answers.
Use a calculator if you disagree.
If you did a problem wrong, work it again.

1. 29×4	**2.** 85×5
3. 93×4	**4.** 52×9
5. 409×5	**6.** 432×8

Math Boxes 9.6

1. There are _____ flowers in $\frac{3}{10}$ of a bunch of 10 flowers.

There are _____ minutes in $\frac{1}{4}$ of an hour.

I have 5 cars. This is $\frac{1}{3}$ of a set of cars. How many cars are in the complete set?

_____ cars

2. How many 4s in 2,000? _______

$2{,}000 \div 4 =$ _______

_______ $\times 4 = 2{,}000$

What number times $7 = 6{,}300$? _______

$6{,}300 \div 7 =$ _______

_______ $\times 7 = 6{,}300$

3. The length of the longer side is _____ units.

The length of the shorter side is _____ units.

The area of the rectangle is _____ square units.

SRB 138

4. Determine the total cost.

4 boxes of cereal at $2.98 each $__________

2 gallons of milk at $3.09 each $__________

Total: $__________

SRB 188 189 191 192

5. Use the partial-products algorithm to solve.

$$\begin{array}{r} 59 \\ \times\ 3 \\ \hline \end{array} \qquad \begin{array}{r} 489 \\ \times\ \ 7 \\ \hline \end{array} \qquad \begin{array}{r} 608 \\ \times\ \ 9 \\ \hline \end{array}$$

SRB 58 59

6. Solve.

$(40 \times 3) \div 2 =$ _______

$4 \times (300 \div 6) =$ _______

$(7 \times 80) + 140 =$ _______

SRB 16 17

Sharing Money

Work with a partner. Put your play money in a bank for both of you to use.

1. If \$54 is shared equally by 3 people, how much does each person get?

   a. How many \$10 bills does each person get? _____ \$10 bill(s)

   b. How many dollars are left to share? \$__________

   c. How many \$1 bills does each person get? _____ \$1 bill(s)

   d. Number model: \$54 ÷ 3 = \$__________

2. If \$204 is shared equally by 6 people, how much does each person get?

   a. How many \$100 bills does each person get? _____ \$100 bill(s)

   b. How many \$10 bills does each person get? _____ \$10 bill(s)

   c. How many dollars are left to share? \$__________

   d. How many \$1 bills does each person get? _____ \$1 bill(s)

   e. Number model: \$204 ÷ 6 = \$__________

3. If \$71 is shared equally by 5 people, how much does each person get?

   a. How many \$10 bills does each person get? _____ \$10 bill(s)

   b. How many dollars are left to share? \$__________

   c. How many \$1 bills does each person get? _____ \$1 bill(s)

   d. How many \$1 bills are left over? _____ \$1 bill(s)

   e. If the leftover \$1 bills are shared equally, how many cents does each person get? _____ ¢

   f. Number model: \$71 ÷ 5 = \$__________

4. \$84 ÷ 3 = \$____________

5. \$75 ÷ 6 = \$____________

6. \$181 ÷ 4 = \$____________

7. \$617 ÷ 5 = \$____________

Date Time

Math Boxes 9.7

1. How many fifths are shaded?

___ fifths

Write the fraction: ___

Write the mixed number: ___

SRB 30

2. Write 3 fractions that are equivalent to $\frac{8}{12}$.

_____ _____ _____

3. Write six factors of 12.

_____ _____ _____

_____ _____ _____

SRB 37

4. What part of this pizza has been eaten?

____

What part is left?

____

SRB 22 23

5. Use the partial-products algorithm to solve.

$$\begin{array}{r} 296 \\ \times \quad 4 \\ \hline \end{array} \qquad \begin{array}{r} 451 \\ \times \quad 5 \\ \hline \end{array} \qquad \begin{array}{r} 183 \\ \times \quad 7 \\ \hline \end{array}$$

SRB 58 59

6. Measure this line segment.

It is about __________ inches long.

It is about _____ centimeters long.

SRB 119–121 125–127

Date Time

Division with Remainders

Solve the problems below. Remember that you will have to decide what the remainder means in order to answer the questions. You may use your calculator, counters, or pictures.

1. Ruth is buying soda for a party. There are 6 cans in a pack. She needs 44 cans.
   How many 6-packs will she buy? _____ 6-packs

2. Paul is buying tickets to the circus. Each ticket costs \$7. He has \$47.
   How many tickets can he buy? _____ tickets

3. Héctor is standing in line for the roller coaster. There are 33 people in line. Each roller coaster car holds 4 people.
   How many cars are needed to hold 33 people? _____ cars

Pretend that the division key on your calculator is broken. Solve the following problems:

4. Regina is building a fence around her dollhouse. She is making each fence post 5 inches tall. The wood she bought is 36 inches long.
   How many fence posts does each piece of wood make? _____ posts

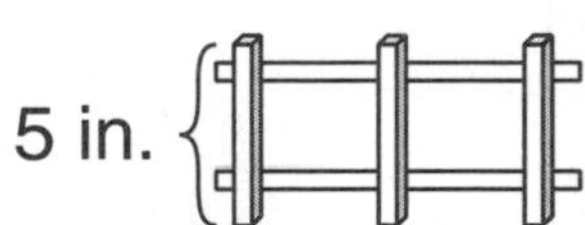

   Explain how you found your answer.

   ____

   ____

5. Missy, Ann, and Herman found a \$10 bill. They want to share the money equally.
   How much money will each person get? __________

   Explain how you found your answer.

   ____

   ____

   ____

Date Time

Using the Partial-Products Algorithm

Multiply. Compare your answers with your partner's answers.
Use a calculator if you disagree.
If you did a problem wrong, work it again.

1. 43×6	**2.** 53×4
3. 68×5	**4.** 74×7
5. 167×4	**6.** 403×3

Date Time

Math Boxes 9.8

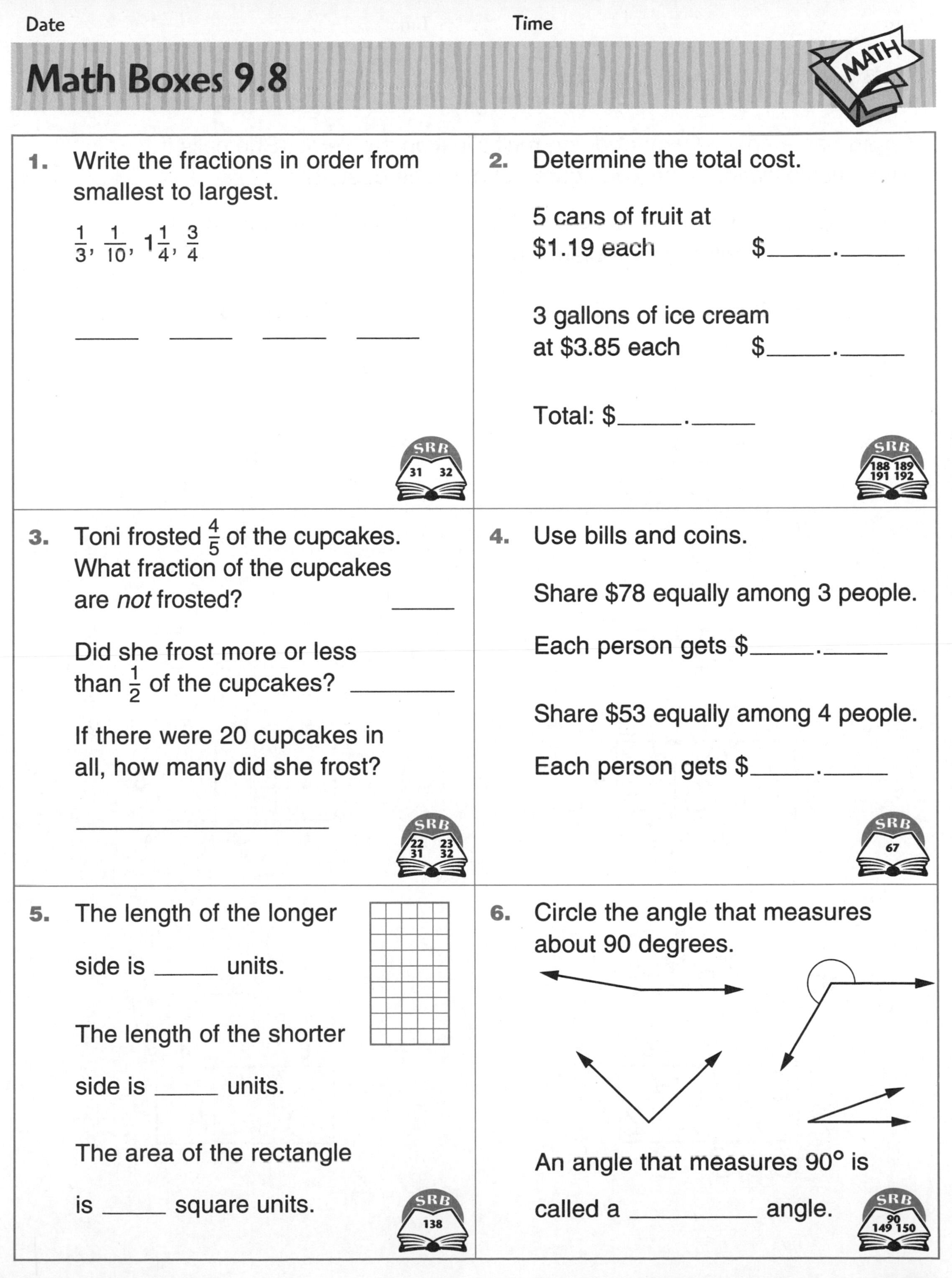

1. Write the fractions in order from smallest to largest.

$\frac{1}{3}$, $\frac{1}{10}$, $1\frac{1}{4}$, $\frac{3}{4}$

_____ _____ _____ _____

2. Determine the total cost.

5 cans of fruit at $1.19 each $_____._____

3 gallons of ice cream at $3.85 each $_____._____

Total: $_____._____

3. Toni frosted $\frac{4}{5}$ of the cupcakes. What fraction of the cupcakes are *not* frosted? _____

Did she frost more or less than $\frac{1}{2}$ of the cupcakes? _______

If there were 20 cupcakes in all, how many did she frost?

____

4. Use bills and coins.

Share $78 equally among 3 people.

Each person gets $_____._____

Share $53 equally among 4 people.

Each person gets $_____._____

5. The length of the longer side is _____ units.

The length of the shorter side is _____ units.

The area of the rectangle is _____ square units.

6. Circle the angle that measures about 90 degrees.

An angle that measures 90° is called a __________ angle.

Lattice Multiplication

Megan has a special way of doing multiplication problems. She calls it lattice multiplication. Can you figure out how she does it?

Study the problems and solutions in Column A. Then try to use lattice multiplication to solve the problems in Column B.

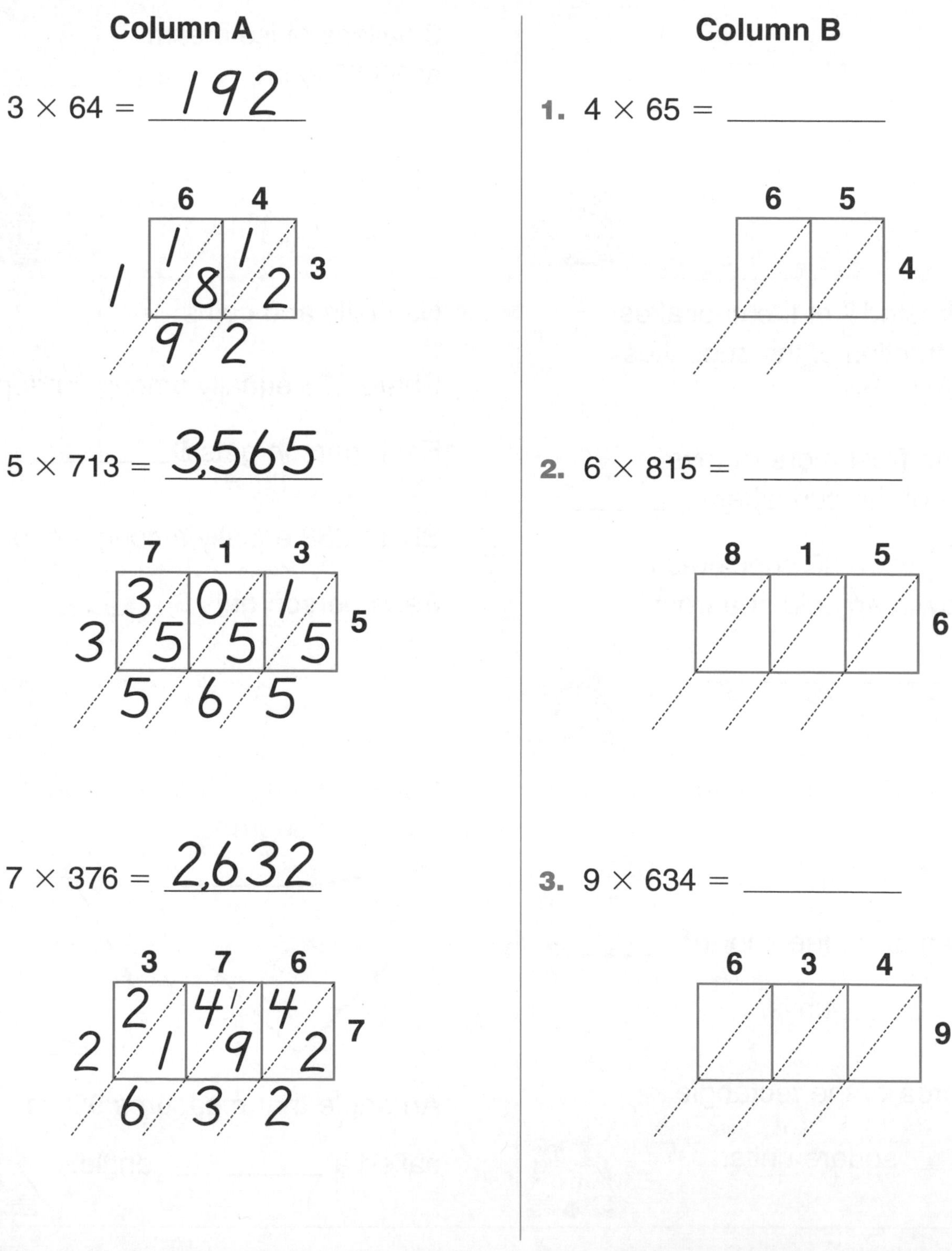

 Use with Lesson 9.9.

Date Time

Lattice Multiplication Practice

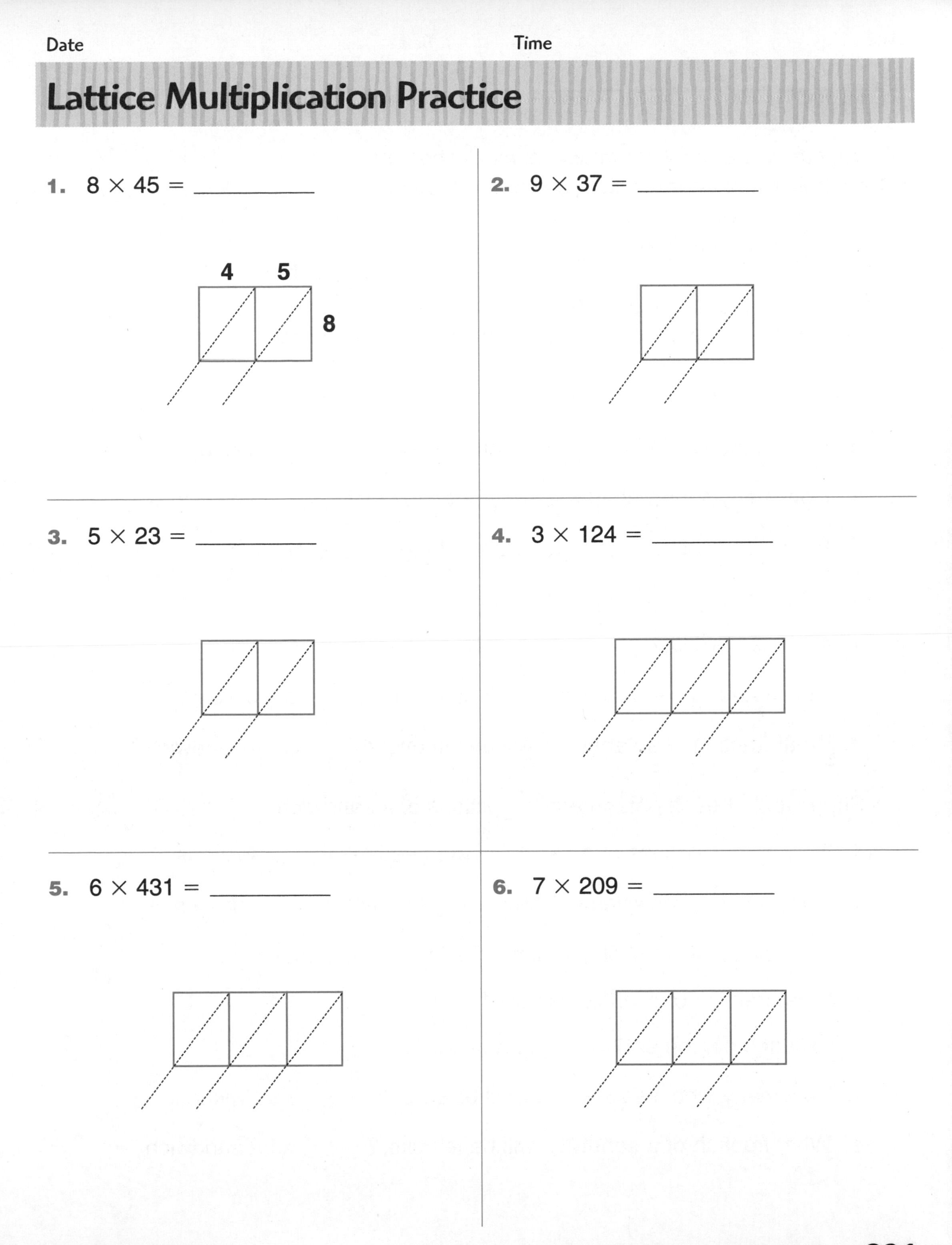

1. $8 \times 45 =$ ________

2. $9 \times 37 =$ ________

3. $5 \times 23 =$ ________

4. $3 \times 124 =$ ________

5. $6 \times 431 =$ ________

6. $7 \times 209 =$ ________

Date Time

Fractions of a Sandwich

Alicia's family is planning a family reunion. There are 20 children and 9 adults. Alicia will order extra-long submarine sandwiches for the reunion. Each sandwich is cut into 6 sections.

1. What is the largest number of family members who might come to the reunion? _____ people

2. Suppose that each person eats 1 section of a sandwich.

   a. How many sections of a sandwich are needed? _____ sections

   b. How many sandwiches will Alicia need to buy? _____ sandwiches

   c. How many sections of a sandwich is that? _____ sections

   d. What fraction of a whole sandwich will each person eat? _____ of a sandwich

   e. How many whole sandwiches will be eaten? _____ sandwiches

   f. What fraction of a sandwich will be left over? _____ of a sandwich

3. Suppose that each person eats 2 sections of a sandwich.

   a. How many sections of a sandwich are needed? _____ sections

   b. How many sandwiches will Alicia need to buy? _____ sandwiches

   c. How many sections of a sandwich is that? _____ sections

   d. What fraction of a whole sandwich will each person eat? _____ of a sandwich

   e. How many whole sandwiches will be eaten? _____ sandwiches

   f. What fraction of a sandwich will be left over? _____ of a sandwich

Date Time

Math Boxes 9.9

1. Fill in the oval next to the best estimate.

1,943 − 488 = ________

0 about 1,000

0 about 1,200

0 about 1,500

0 about 1,800

SRB 167

2. How many 4s in 40? ________

How many 4s in 400? ________

How many 4s in 4,000? ________

How many 10s in 400? ________

How many 100s in 40,000? ________

SRB 18 19

3. Write six factors of 20.

____ ____ ____

____ ____ ____

SRB 37

4. Allison has 58 stickers. She wants to share them among 8 friends.

How many stickers does each friend get?

____

How many stickers are left over?

____

SRB 67

5. Use the partial-products algorithm to solve.

$$\begin{array}{r} 238 \\ \times \quad 6 \\ \hline \end{array} \qquad \begin{array}{r} 574 \\ \times \quad 5 \\ \hline \end{array} \qquad \begin{array}{r} 706 \\ \times \quad 7 \\ \hline \end{array}$$

SRB 58 59

6. Put in the parentheses needed to complete the number models.

$15 + 80 \times 90 = 7{,}215$

$14 - 6 \times 800 = 6{,}400$

$60 \times 79 + 1 = 4{,}800$

SRB 16 17

Date Time

Array Multiplication 2

1. How many squares are in a 20-by-13 array?

Total squares = ______

$20 \times 13 =$ ______

2. How many squares are in an 18-by-30 array?

Total squares = ______

$18 \times 30 =$ ______

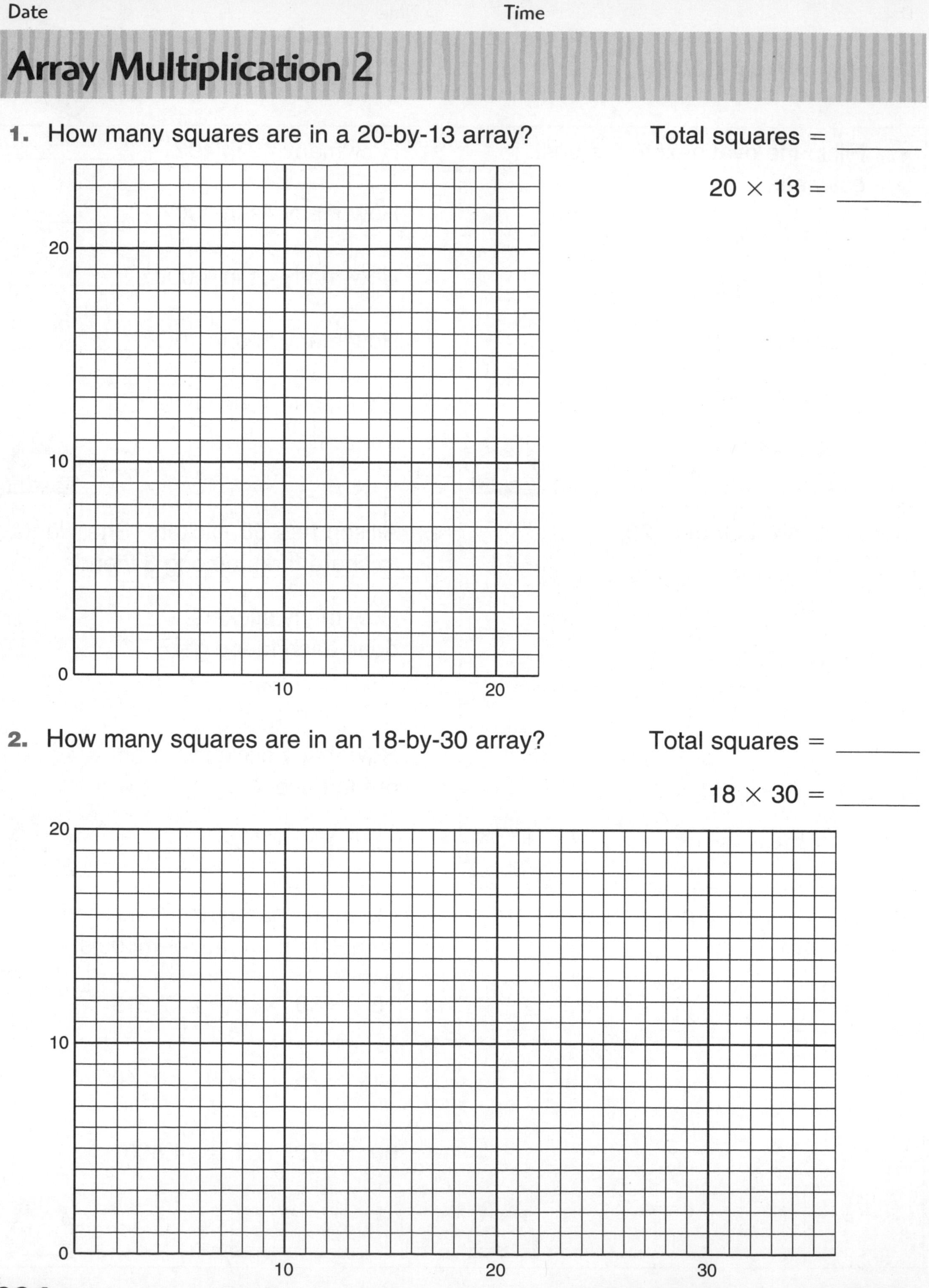

Use with Lessons 9.10 and 9.11.

Date Time

Array Multiplication 3

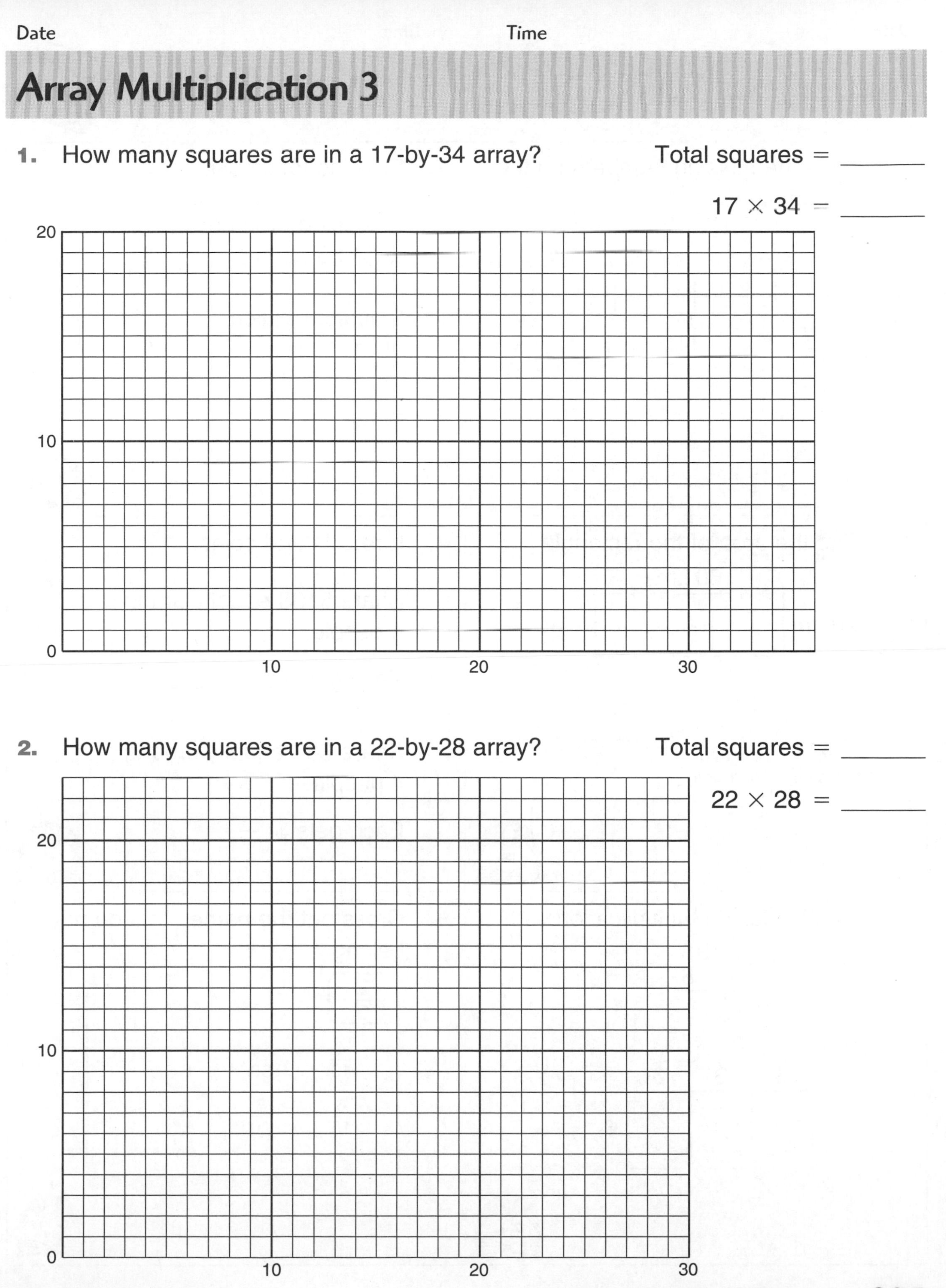

1. How many squares are in a 17-by-34 array?

Total squares = ______

$17 \times 34 =$ ______

2. How many squares are in a 22-by-28 array?

Total squares = ______

$22 \times 28 =$ ______

Date Time

Math Boxes 9.10

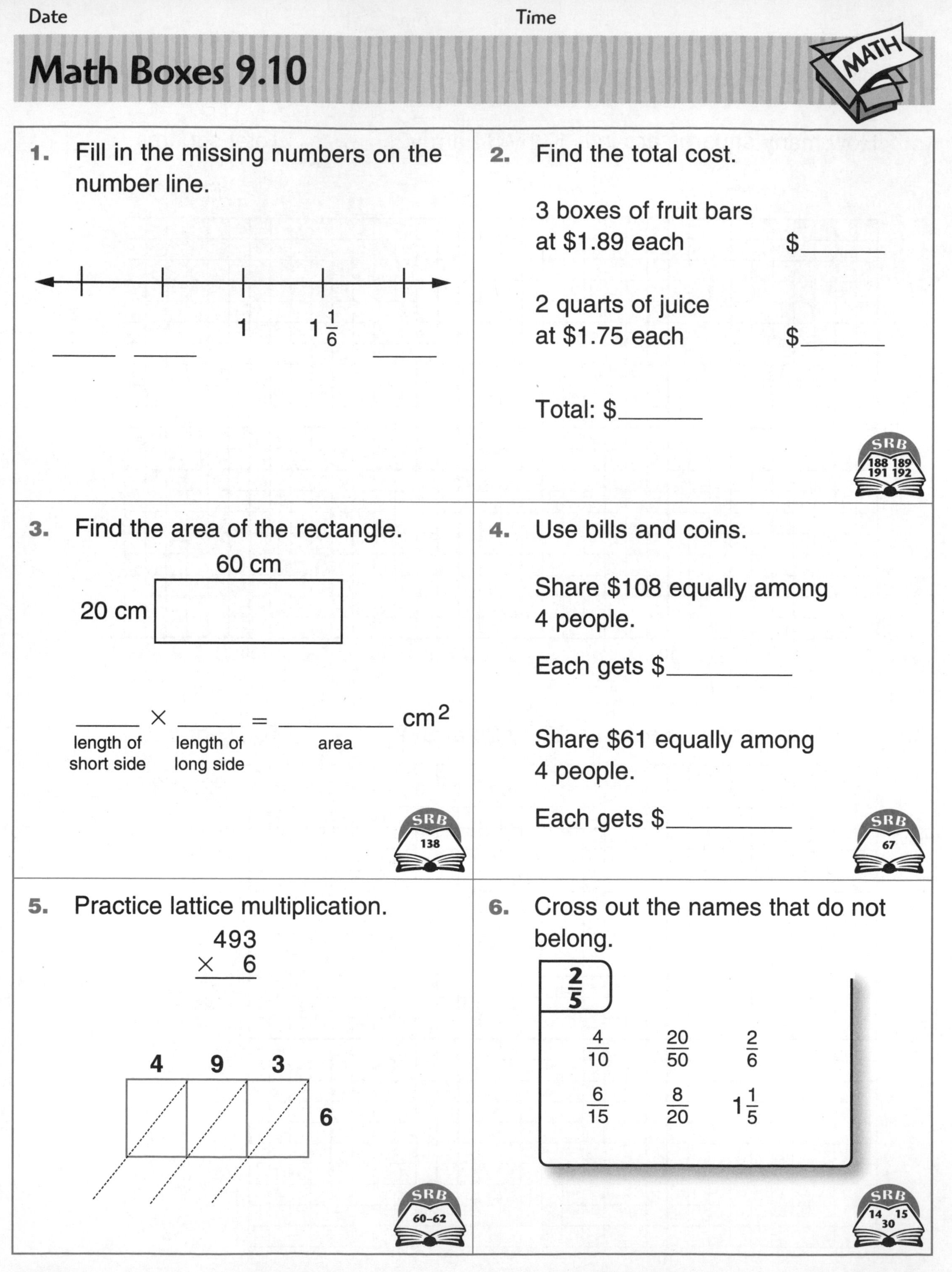

1. Fill in the missing numbers on the number line.

   ____ ____ 1 $1\frac{1}{6}$ ____

2. Find the total cost.

   3 boxes of fruit bars at \$1.89 each \$______

   2 quarts of juice at \$1.75 each \$______

   Total: \$______

   SRB 188 189 191 192

3. Find the area of the rectangle.

   60 cm

   20 cm

   ____ × ____ = ________ cm^2

   length of short side, length of long side, area

   SRB 138

4. Use bills and coins.

   Share \$108 equally among 4 people.

   Each gets \$________

   Share \$61 equally among 4 people.

   Each gets \$________

   SRB 67

5. Practice lattice multiplication.

$$\begin{array}{r} 493 \\ \times \quad 6 \\ \hline \end{array}$$

   4 9 3

   6

   SRB 60–62

6. Cross out the names that do not belong.

   $\frac{2}{5}$

   $\frac{4}{10}$ $\frac{20}{50}$ $\frac{2}{6}$

   $\frac{6}{15}$ $\frac{8}{20}$ $1\frac{1}{5}$

   SRB 14 15 30

Date Time

Multiplication with Multiples of 10

Multiply. Compare your answers with your partner's answer.
Use a calculator if you disagree.
If you did a problem wrong, work it again.

Example

$$\begin{array}{r} 30 \\ \times\ 26 \\ \hline \end{array}$$

20[30s]→ 600
6[30s]→ + 180
780

1. $$\begin{array}{r} 70 \\ \times\ 18 \\ \hline \end{array}$$

2. $$\begin{array}{r} 88 \\ \times\ 40 \\ \hline \end{array}$$

3. $$\begin{array}{r} 60 \\ \times\ 35 \\ \hline \end{array}$$

4. $$\begin{array}{r} 80 \\ \times\ 44 \\ \hline \end{array}$$

5. $$\begin{array}{r} 90 \\ \times\ 63 \\ \hline \end{array}$$

Date Time

Math Boxes 9.11

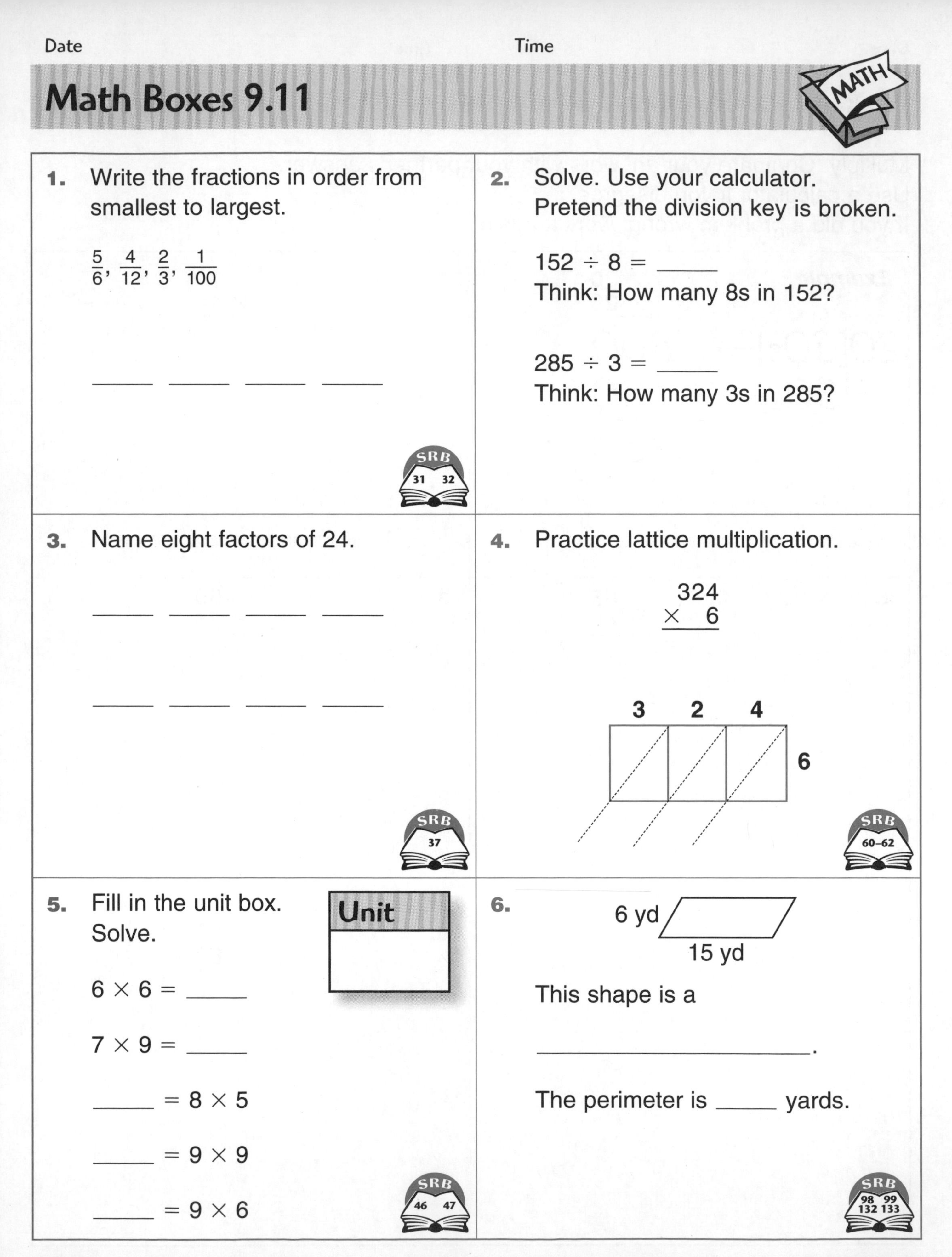

1. Write the fractions in order from smallest to largest.

$\frac{5}{6}, \frac{4}{12}, \frac{2}{3}, \frac{1}{100}$

_____ _____ _____ _____

SRB 31 32

2. Solve. Use your calculator. Pretend the division key is broken.

$152 \div 8 =$ _____
Think: How many 8s in 152?

$285 \div 3 =$ _____
Think: How many 3s in 285?

3. Name eight factors of 24.

_____ _____ _____ _____

_____ _____ _____ _____

SRB 37

4. Practice lattice multiplication.

$$\begin{array}{r} 324 \\ \times \quad 6 \\ \hline \end{array}$$

SRB 60–62

5. Fill in the unit box. Solve.

Unit

$6 \times 6 =$ _____

$7 \times 9 =$ _____

_____ $= 8 \times 5$

_____ $= 9 \times 9$

_____ $= 9 \times 6$

SRB 46 47

6.

This shape is a

_____________________.

The perimeter is _____ yards.

SRB 98 99 132 133

Date Time

2-Digit Multiplication

Multiply using the partial-products algorithm. Compare your answers with your partner's answers. Use a calculator if you disagree.
If you did a problem wrong, work it again.

1. 24×16	**2.** 42×31	**3.** 12×87
4. 59×79	**5.** 36×14	**6.** 42×53
7. 23×81	**8.** 63×12	**9.** 49×38

Date Time

Accurate Measures

Use fractions to carefully measure these drawings using both the inch and centimeter sides of your ruler.

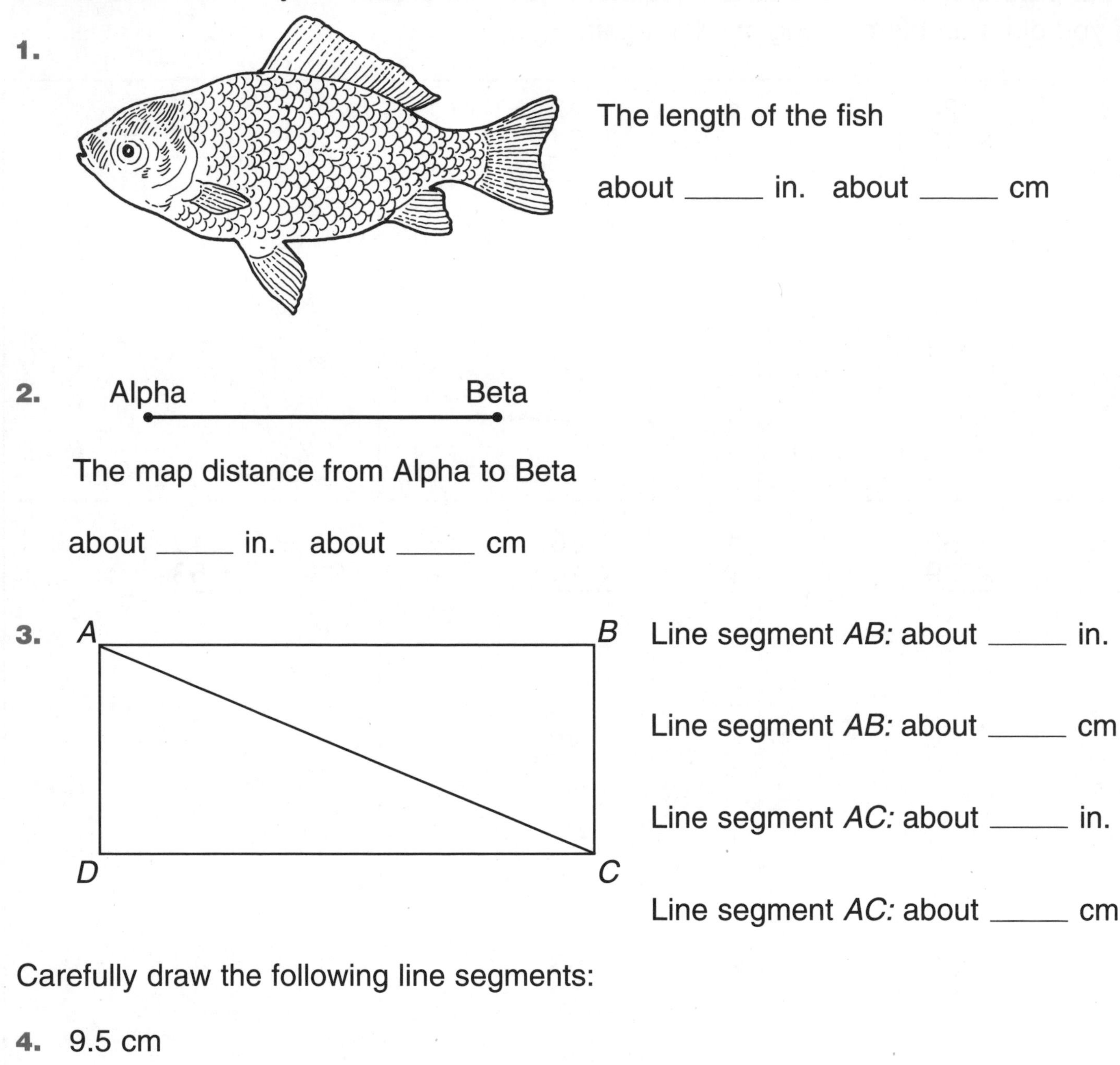

1. The length of the fish

about _____ in. about _____ cm

2. Alpha Beta

The map distance from Alpha to Beta

about _____ in. about _____ cm

3. A B C D

Line segment *AB:* about _____ in.

Line segment *AB:* about _____ cm

Line segment *AC:* about _____ in.

Line segment *AC:* about _____ cm

Carefully draw the following line segments:

4. 9.5 cm

5. $4\frac{1}{4}$ in.

6. 2 cm shorter than 9.5 cm

7. $1\frac{1}{4}$ in. shorter than $4\frac{1}{4}$ in.

Date Time

Math Boxes 9.12

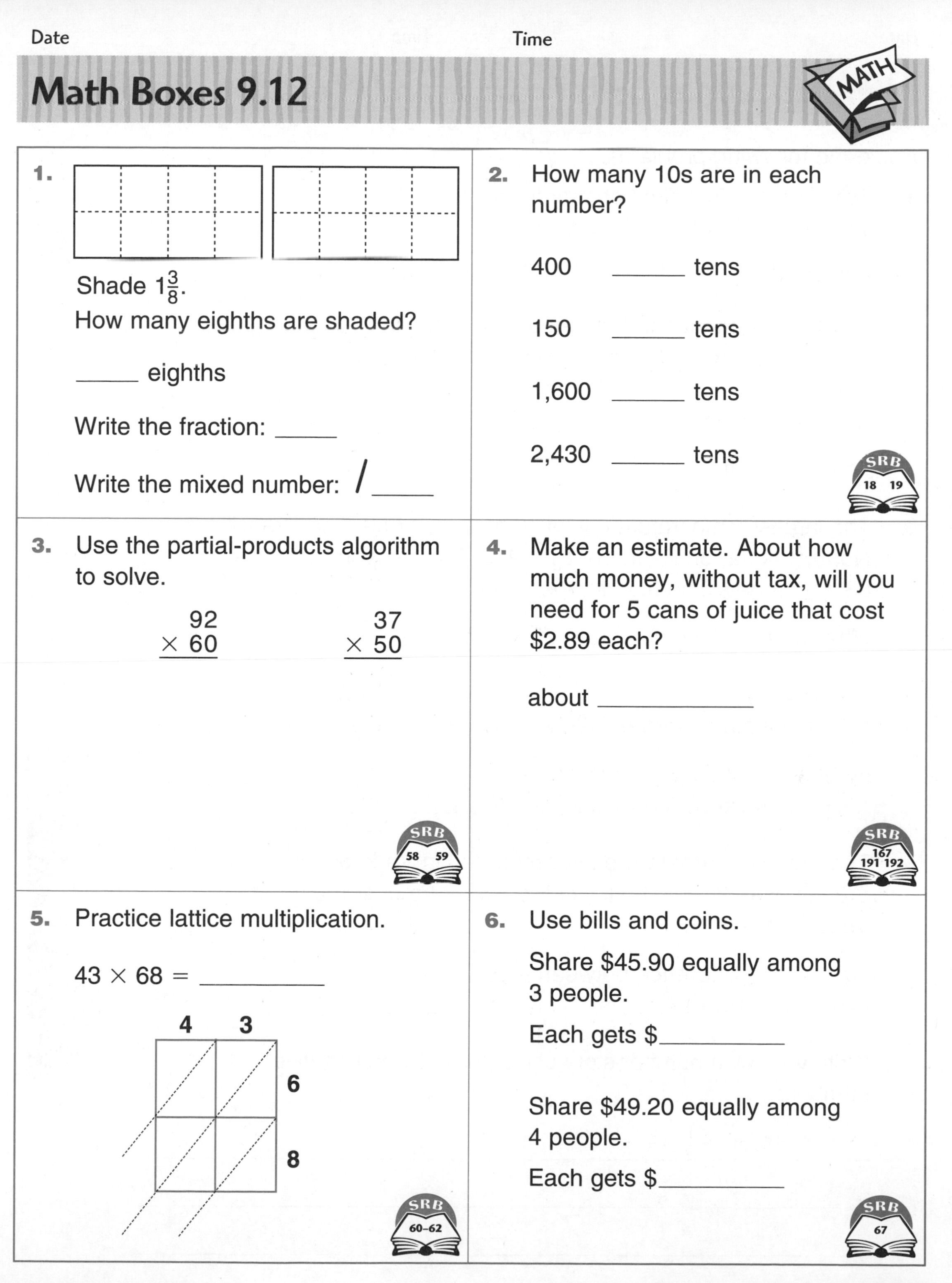

1. Shade $1\frac{3}{8}$.

How many eighths are shaded?

_____ eighths

Write the fraction: _____

Write the mixed number: / _____

2. How many 10s are in each number?

400 _____ tens

150 _____ tens

1,600 _____ tens

2,430 _____ tens

SRB 18 19

3. Use the partial-products algorithm to solve.

$$\begin{array}{r} 92 \\ \times\ 60 \\ \hline \end{array} \qquad \begin{array}{r} 37 \\ \times\ 50 \\ \hline \end{array}$$

SRB 58 59

4. Make an estimate. About how much money, without tax, will you need for 5 cans of juice that cost \$2.89 each?

about __________

SRB 167 191 192

5. Practice lattice multiplication.

$43 \times 68 =$ __________

SRB 60–62

6. Use bills and coins.

Share \$45.90 equally among 3 people.

Each gets \$__________

Share \$49.20 equally among 4 people.

Each gets \$__________

SRB 67

Date Time

Number Stories with Positive and Negative Numbers

Solve the following problems.

1. Jim records his weight change weekly. This week he recorded −3 pounds.

   Can you tell how much he weighs? ______

2. The largest change in temperature in a single day took place in January 1916 in Browning, Montana. The temperature dropped 100°F that day. The temperature was 44°F when it started dropping.

   How low did it go? ______

3. The largest temperature rise in 12 hours took place in Granville, North Dakota, on February 21, 1918. The temperature rose 83°F that day. The high temperature was 50°F.

   What was the low temperature? ______

4. On January 12, 1911, the temperature in Rapid City, South Dakota, fell from 49°F at 6 A.M. to –13°F at 8 A.M.

   By how many degrees did the temperature drop in those 2 hours? ______

5. The highest temperature ever recorded in Verkhoyansk, Siberia, was 98°F. The lowest temperature ever recorded there was –94°F.

   What is the difference between those two temperatures? ______

6. Write your own number story using positive and negative numbers.

   ______

   ______

   ______

°F
220
210
200
190
180
170
160
150
140
130
120
110
100
90
80
70
60
50
40
30
20
10
0
–10
–20
–30
–40

Date Time

Multiplication Strategies

Try using your favorite strategy to solve each problem. Compare your answers with your partner's answers. Use a calculator if you disagree. If you make a mistake, solve the problem again.

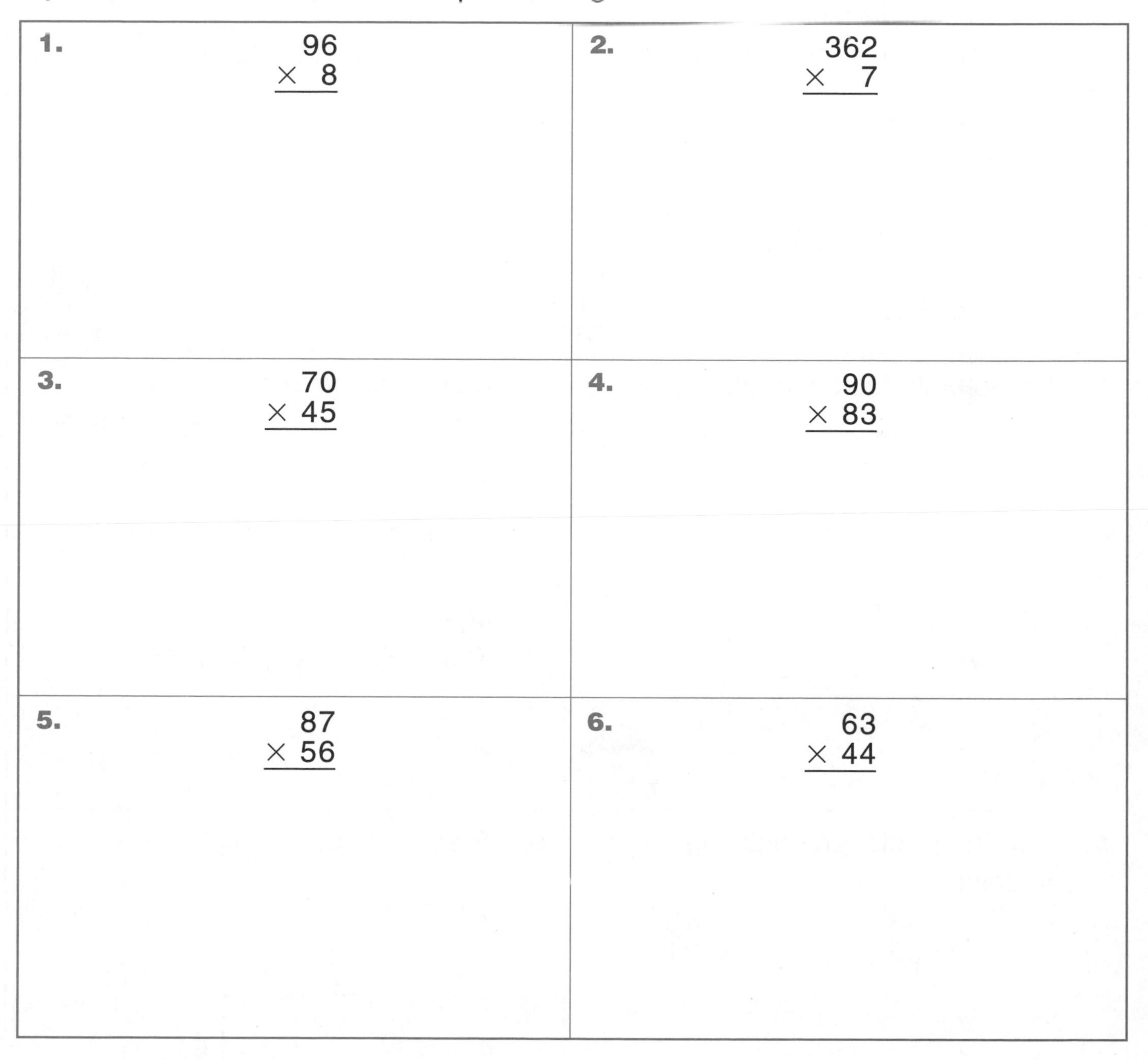

1. 96×8	**2.** 362×7
3. 70×45	**4.** 90×83
5. 87×56	**6.** 63×44

7. My favorite multiplication strategy is ______________________.

I like this strategy best because ______________________

____

Math Boxes 9.13

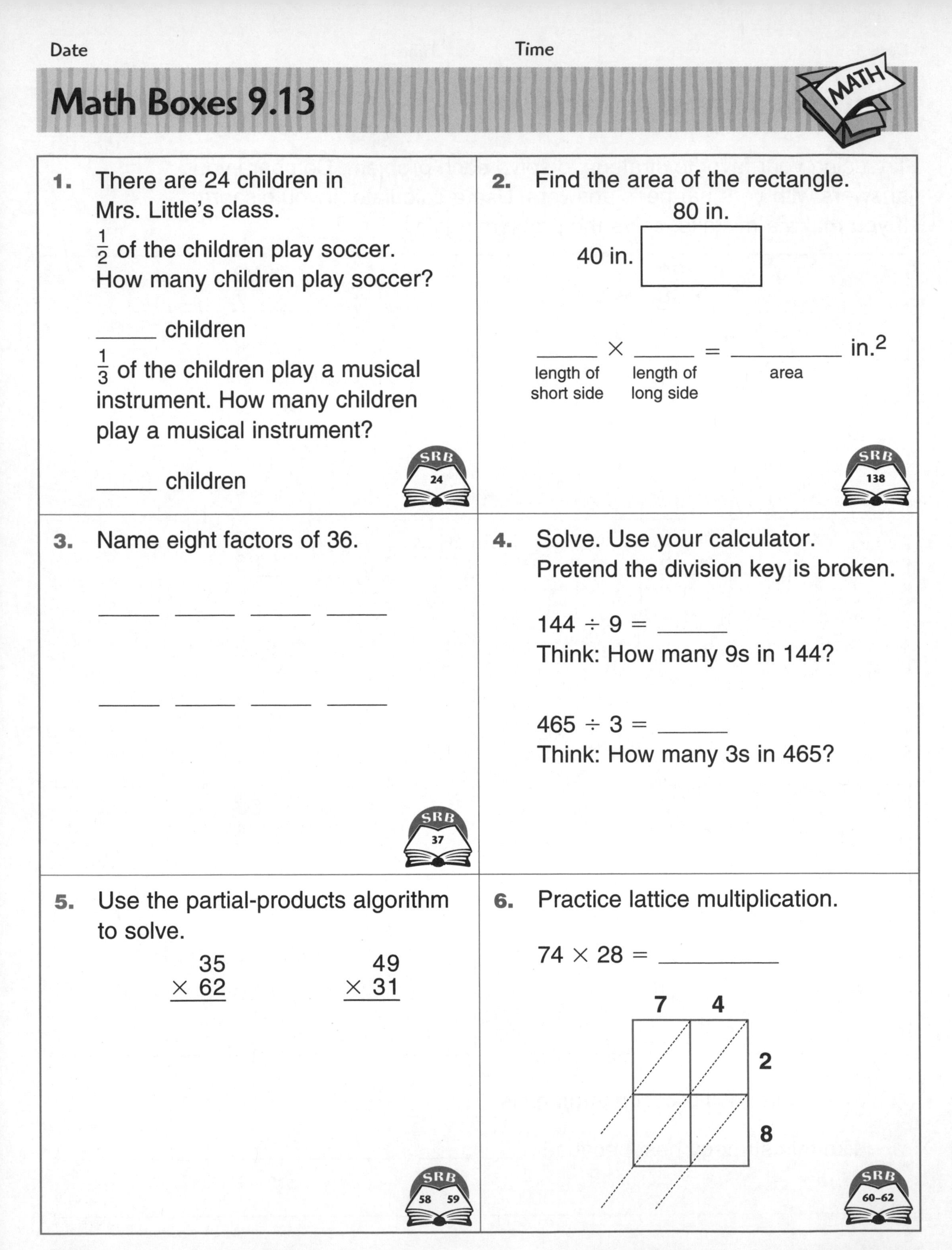

1. There are 24 children in Mrs. Little's class.
$\frac{1}{2}$ of the children play soccer.
How many children play soccer?

_____ children

$\frac{1}{3}$ of the children play a musical instrument. How many children play a musical instrument?

_____ children

SRB 24

2. Find the area of the rectangle.

80 in.

40 in.

_____ × _____ = _________ in.^{2}

length of short side, length of long side, area

SRB 138

3. Name eight factors of 36.

_____ _____ _____ _____

_____ _____ _____ _____

SRB 37

4. Solve. Use your calculator. Pretend the division key is broken.

$144 \div 9 =$ _____
Think: How many 9s in 144?

$465 \div 3 =$ _____
Think: How many 3s in 465?

5. Use the partial-products algorithm to solve.

$$\begin{array}{r} 35 \\ \times\ 62 \\ \hline \end{array} \qquad \begin{array}{r} 49 \\ \times\ 31 \\ \hline \end{array}$$

SRB 58 59

6. Practice lattice multiplication.

$74 \times 28 =$ _________

SRB 60–62

Date Time

Math Boxes 9.14

1. Measure this line segment.

It is about _____ inches long.

Draw a line segment $1\frac{3}{4}$ inches long.

SRB 125–127

2. Measure this line segment.

It is about _____ centimeters long.

Draw a line segment 3.5 centimeters long.

SRB 119–121

3. Circle the most appropriate unit.

length of calculator:

inches feet miles

weight of an adult:

ounces pounds tons

amount of gas in car:

cups pints gallons

4. Solve.

1 foot = _____ inches

_____ feet = 36 inches

1 yard = _____ feet

_____ yards = 15 feet

1 yard = _____ inches

SRB 128 129

5. Find the median of the following numbers.

34, 56, 34, 16, 33, 27, 45

Median: _____

SRB 74

6. Circle the tool you would use to find

length of a pen:

ruler compass scale

weight of a dime:

ruler compass scale

way to get home:

ruler compass scale

SRB 144–146

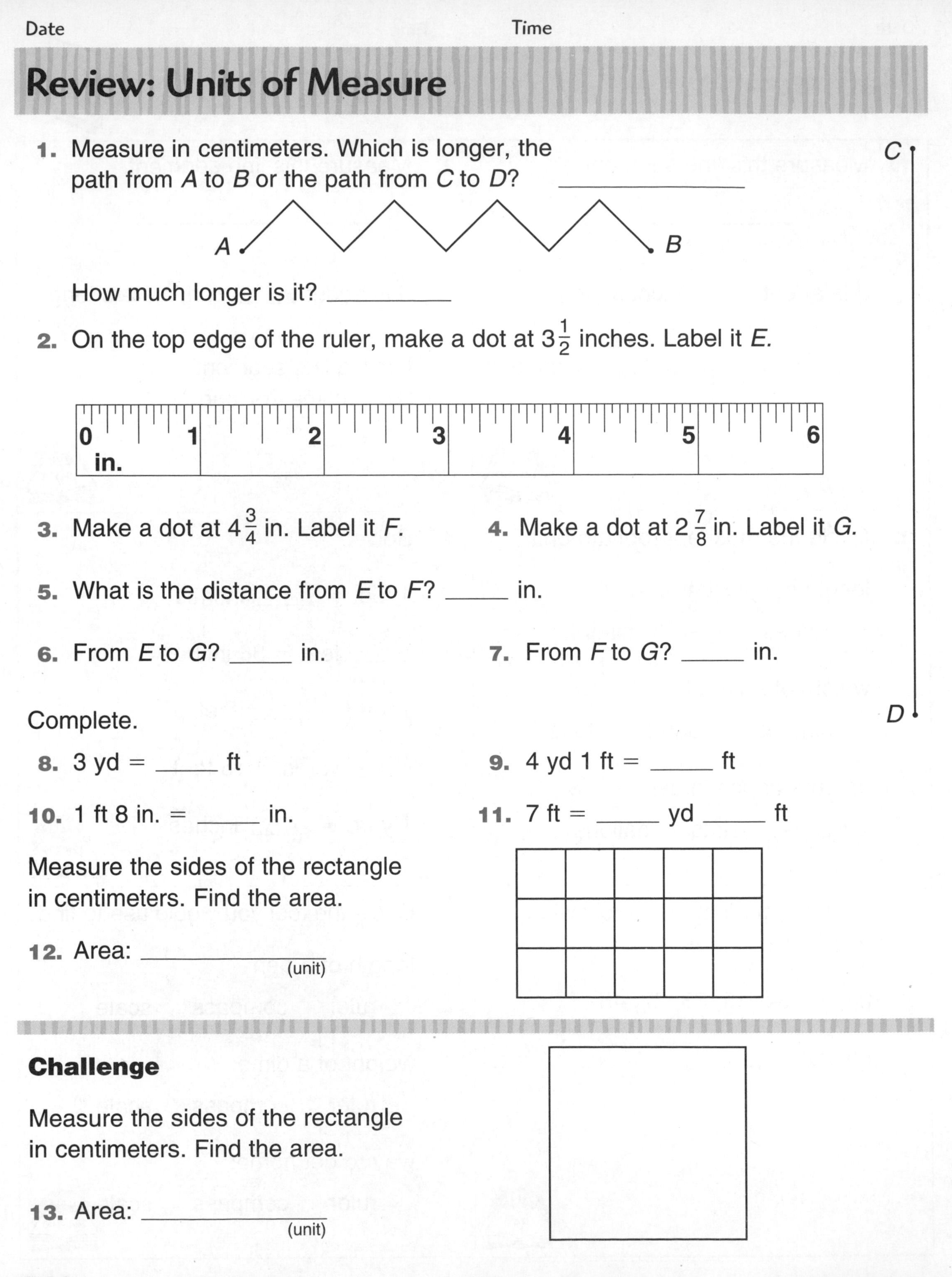

Date Time

Review: Units of Measure

1. Measure in centimeters. Which is longer, the path from *A* to *B* or the path from *C* to *D*? ____________

How much longer is it? ____________

2. On the top edge of the ruler, make a dot at $3\frac{1}{2}$ inches. Label it *E*.

3. Make a dot at $4\frac{3}{4}$ in. Label it *F*.

4. Make a dot at $2\frac{7}{8}$ in. Label it *G*.

5. What is the distance from *E* to *F*? _____ in.

6. From *E* to *G*? _____ in.

7. From *F* to *G*? _____ in.

Complete.

8. 3 yd = _____ ft

9. 4 yd 1 ft = _____ ft

10. 1 ft 8 in. = _____ in.

11. 7 ft = _____ yd _____ ft

Measure the sides of the rectangle in centimeters. Find the area.

12. Area: ______________ (unit)

Challenge

Measure the sides of the rectangle in centimeters. Find the area.

13. Area: ______________ (unit)

Date Time

Earth Layers

Earth is made of layers. The outer layer (the part you stand on) is called the crust. Compared to the other layers, the crust is very thin—it ranges from 8 to 80 kilometers in depth. If Earth were a huge egg, the crust would not be much thicker than the shell.

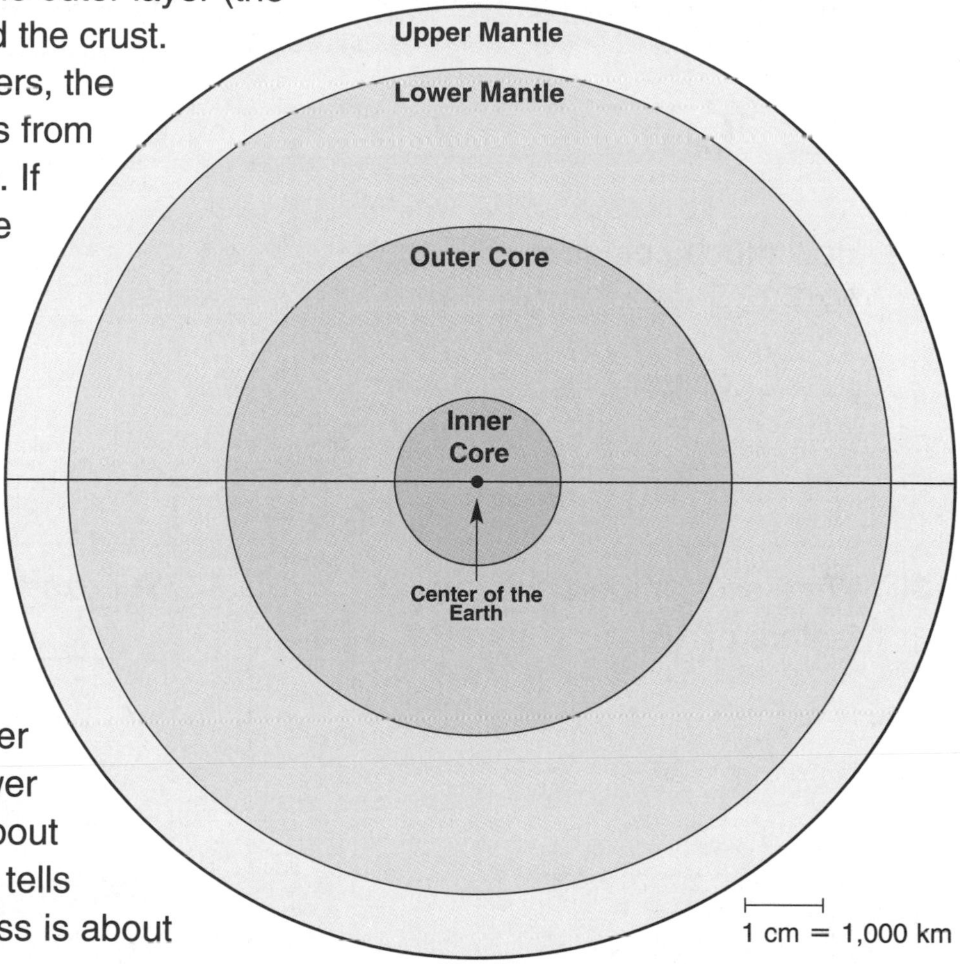

The picture is a scale drawing of Earth's layers. In the drawing, each centimeter stands for 1,000 kilometers. You can estimate the actual thickness of each layer by measuring with a centimeter ruler. For example, the lower mantle in the drawing is about 2.1 centimeters wide. This tells you that the actual thickness is about 2,100 kilometers.

1. Imagine that you could dig a hole to the center of Earth. About how deep would the hole be? about __________ kilometers

2. What is the diameter of Earth? about __________ kilometers

3. The closer you get to the center of Earth, the hotter it gets. Study the data in the table. In which layer is the temperature about 800°C?

____

4. In which layer is the temperature about 3,000°C?

____

Temperatures Inside Earth	
50 km below the surface	about 800°C
1,000 km below the surface	about 1,800°C
2,000 km below the surface	about 3,000°C
3,200 km below the surface	about 3,500°C
at the center of Earth	about 4,000°C

Date Time

Math Boxes 10.1

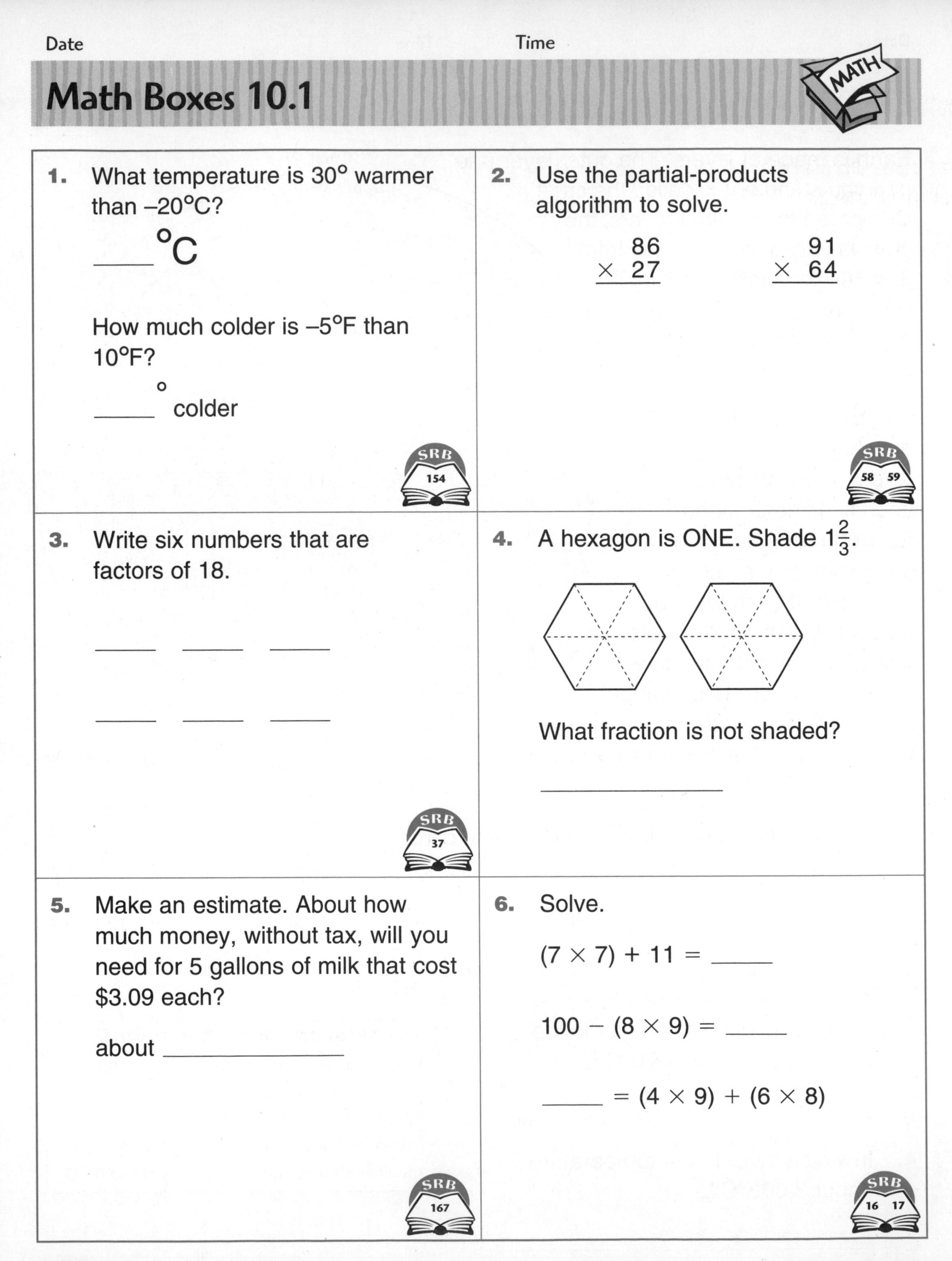

1. What temperature is 30° warmer than –20°C?

   ____ °C

   How much colder is –5°F than 10°F?

   ____ ° colder

   SRB 154

2. Use the partial-products algorithm to solve.

   $$\begin{array}{r} 86 \\ \times\ 27 \\ \hline \end{array} \qquad \begin{array}{r} 91 \\ \times\ 64 \\ \hline \end{array}$$

   SRB 58 59

3. Write six numbers that are factors of 18.

   ____ ____ ____

   ____ ____ ____

   SRB 37

4. A hexagon is ONE. Shade $1\frac{2}{3}$.

   What fraction is not shaded?

   ____

5. Make an estimate. About how much money, without tax, will you need for 5 gallons of milk that cost $3.09 each?

   about ________

   SRB 167

6. Solve.

   $(7 \times 7) + 11 =$ ____

   $100 - (8 \times 9) =$ ____

   ____ $= (4 \times 9) + (6 \times 8)$

   SRB 16 17

Use with Lesson 10.1.

Date Time

Volume of Boxes

Part 1 Use the patterns on *Math Masters,* page 166 to build Boxes A, B, C, and D. Record the results in the table.

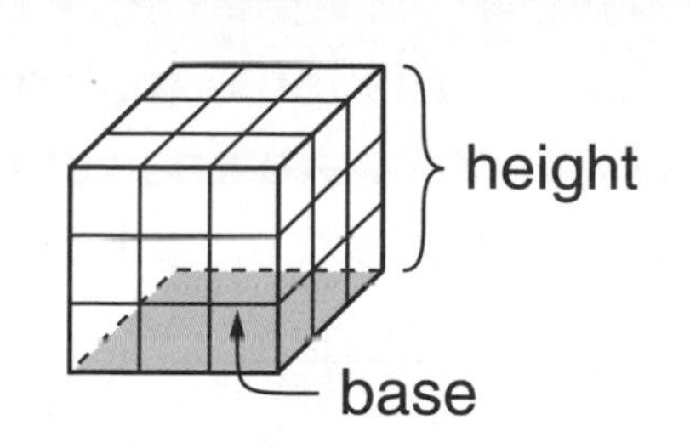

Box	Number of cm Cubes		Area of Base	Height	Volume
	Estimate	Exact	(square cm)	(cm)	(cubic cm)
A					
B					
C					
D					

Part 2 The following patterns are for Boxes E, F, and G. Each square stands for 1 square centimeter. Find the volume of each box. (Do not cut out the patterns.)

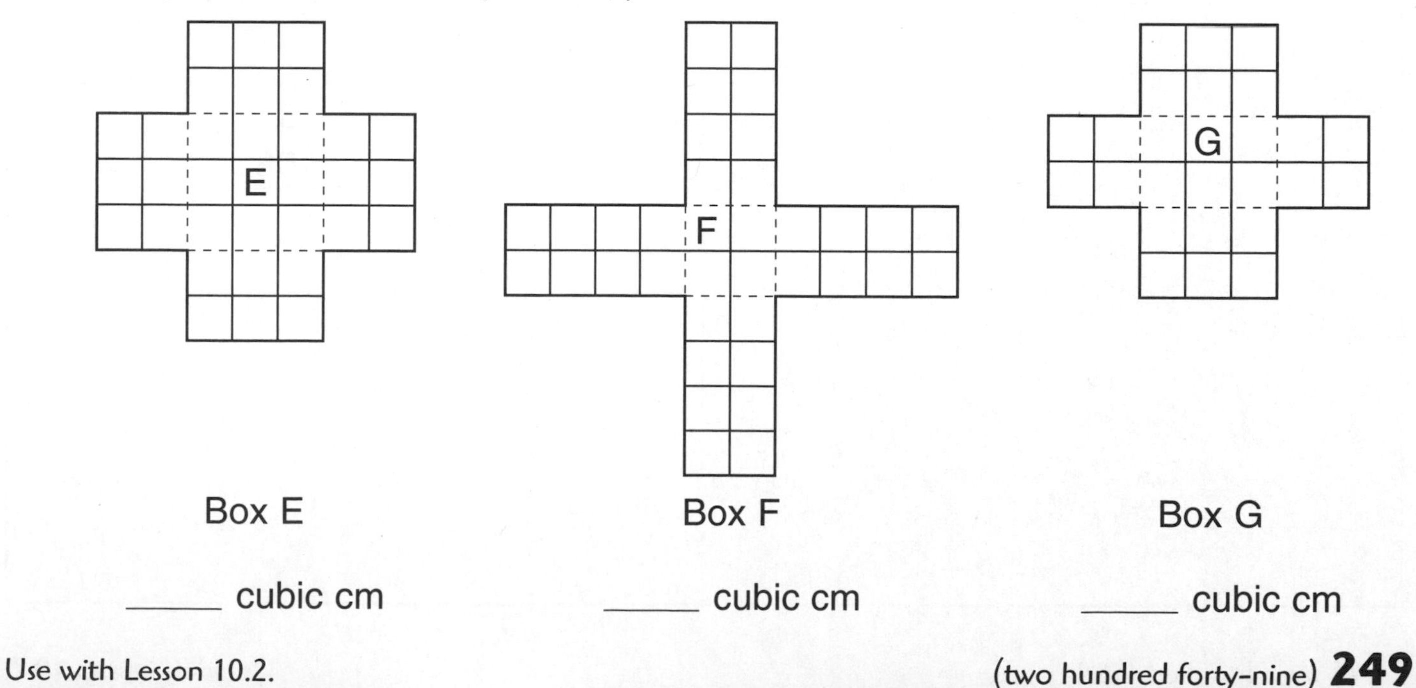

Date Time

Multiplication Practice

Use your favorite multiplication algorithm to solve the following problems. Then, compare answers with your partner. Use a calculator if you disagree. If you made a mistake on a problem, try to solve it again.

1. 427×3	**2.** 505×8
3. 20×90	**4.** 67×40
5. 74×35	**6.** 37×58

Use with Lesson 10.2.

Date Time

Weight and Volume

Complete Parts 1 and 2 before the start of Lesson 10.5.

Part 1 Try to order the objects on display from heaviest to lightest. Lift them to help you guess. Record your guesses below.

Names of objects in order

heaviest ____________

2nd heaviest ____________

3rd heaviest ____________

lightest ____________

Part 2 Try to order the objects on display from largest to smallest volume. Record your guesses below.

Names of objects in order

largest ____________

2nd largest ____________

3rd largest ____________

smallest ____________

Complete Parts 3 and 4 as part of Lesson 10.5.

Part 3 Record the actual order of the objects from heaviest to lightest. Were your guesses correct?

Names of objects in order

heaviest ____________

2nd heaviest ____________

3rd heaviest ____________

lightest ____________

Part 4 Record the actual order of the objects from largest to smallest volume. Were your guesses correct?

Names of objects in order

largest ____________

2nd largest ____________

3rd largest ____________

smallest ____________

Review Complete the number models.

1. $(3 \times 8) - 7 =$ _____

$3 \times (8 - 7) =$ _____

2. _____ $= (18 \div 2) + 4$

_____ $= 18 \div (2 + 4)$

3. $(15 + 25) - 8 =$ _____

$15 + (25 - 8) =$ _____

4. _____ $= (6 + 4) \times (6 - 4)$

_____ $= 6 + (4 \times 6) - 4$

5. $37 - (12 - 5) =$ _____

$(37 - 12) - 5 =$ _____

6. _____ $= (24 \div 4) \div 2$

_____ $= 24 \div (4 \div 2)$

Date Time

Math Boxes 10.2

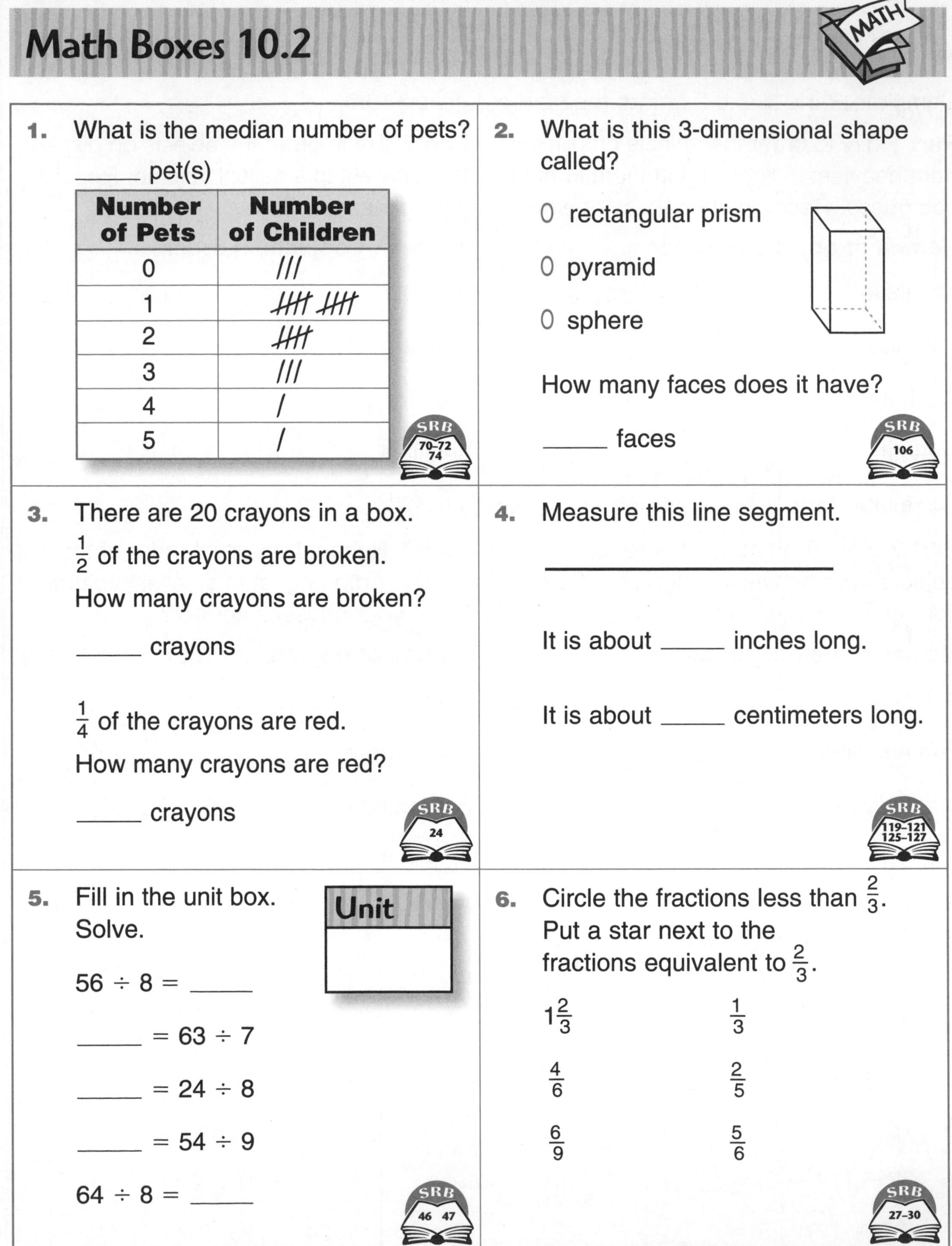

1. What is the median number of pets?

_____ pet(s)

Number of Pets	Number of Children
0	///
1	𝍸 𝍸
2	𝍸
3	///
4	/
5	/

SRB 70–72 74

2. What is this 3-dimensional shape called?

0 rectangular prism

0 pyramid

0 sphere

How many faces does it have?

_____ faces

SRB 106

3. There are 20 crayons in a box.

$\frac{1}{2}$ of the crayons are broken.

How many crayons are broken?

_____ crayons

$\frac{1}{4}$ of the crayons are red.

How many crayons are red?

_____ crayons

SRB 24

4. Measure this line segment.

It is about _____ inches long.

It is about _____ centimeters long.

SRB 119–121 125–127

5. Fill in the unit box. Solve.

Unit

$56 \div 8 =$ _____

_____ $= 63 \div 7$

_____ $= 24 \div 8$

_____ $= 54 \div 9$

$64 \div 8 =$ _____

SRB 46 47

6. Circle the fractions less than $\frac{2}{3}$. Put a star next to the fractions equivalent to $\frac{2}{3}$.

$1\frac{2}{3}$ $\frac{1}{3}$

$\frac{4}{6}$ $\frac{2}{5}$

$\frac{6}{9}$ $\frac{5}{6}$

SRB 27–30

Date Time

Body Measures

Work with a partner to make each measurement to the nearest $\frac{1}{4}$ inch.

	Adult at Home	Me (Now)	Me (Later)
Date			
height	about ______ in.	about ______ in.	about ______ in.
shoe length	about ______ in.	about ______ in.	about ______ in.
around neck	about ______ in.	about ______ in.	about ______ in.
around wrist	about ______ in.	about ______ in.	about ______ in.
waist to floor	about ______ in.	about ______ in.	about ______ in.
forearm	about ______ in.	about ______ in.	about ______ in.
hand span	about ______ in.	about ______ in.	about ______ in.
arm span	about ______ in.	about ______ in.	about ______ in.
______________	about ______ in.	about ______ in.	about ______ in.
______________	about ______ in.	about ______ in.	about ______ in.

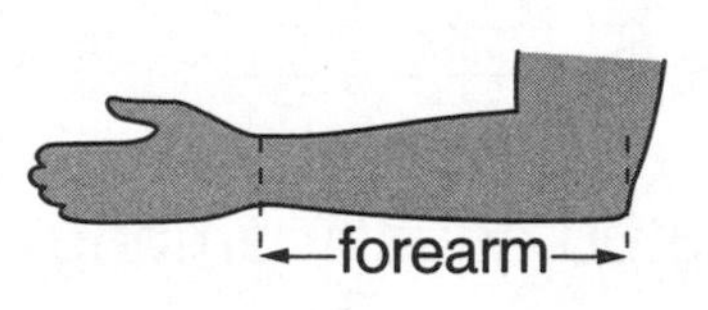

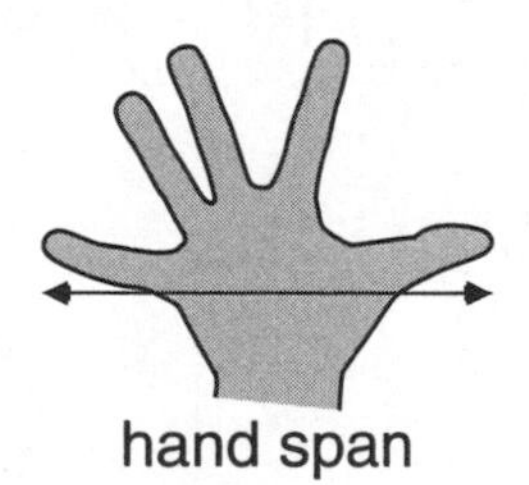
hand span

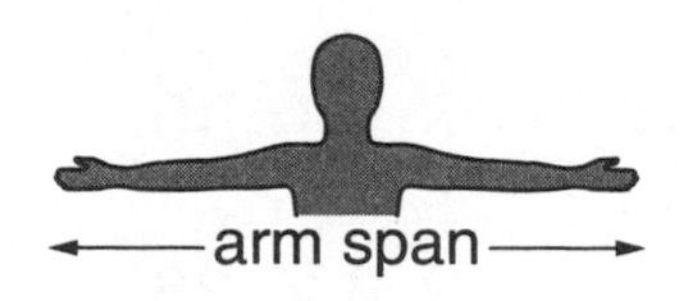

Date Time

Math Boxes 10.3

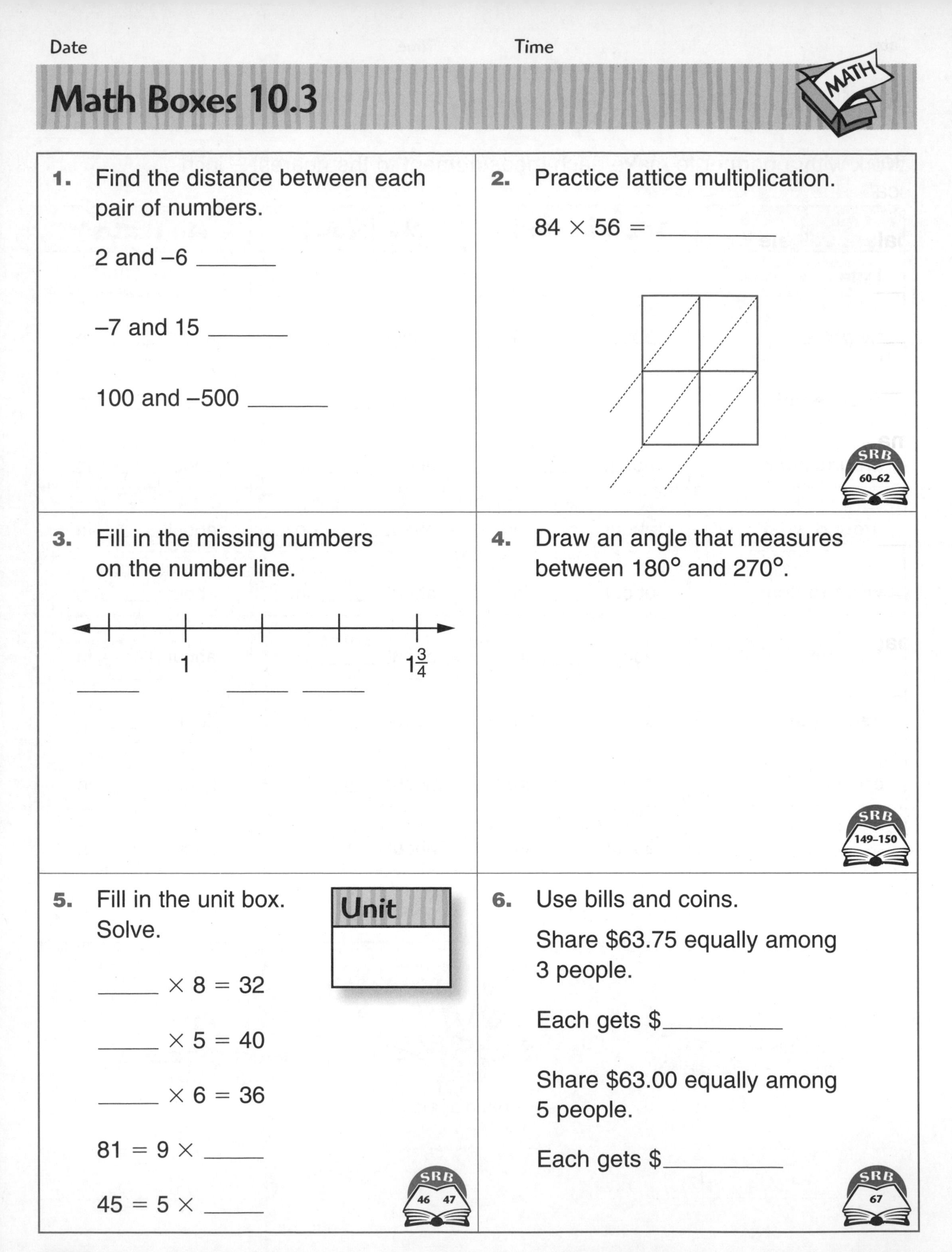

1. Find the distance between each pair of numbers.

   2 and –6 ______

   –7 and 15 ______

   100 and –500 ______

2. Practice lattice multiplication.

   $84 \times 56 =$ ________

   SRB 60–62

3. Fill in the missing numbers on the number line.

   ______ 1 ______ ______ $1\frac{3}{4}$

4. Draw an angle that measures between 180° and 270°.

   SRB 149–150

5. Fill in the unit box. Solve.

   Unit

   ______ $\times 8 = 32$

   ______ $\times 5 = 40$

   ______ $\times 6 = 36$

   $81 = 9 \times$ ______

   $45 = 5 \times$ ______

   SRB 46 47

6. Use bills and coins.

   Share $63.75 equally among 3 people.

   Each gets $________

   Share $63.00 equally among 5 people.

   Each gets $________

   SRB 67

Date Time

Scales

Refer to pages 147 and 148 in your *Student Reference Book.* For each scale shown, list three things you could weigh on the scale.

balance scale

____

____

____

produce scale

____

____

____

market scale

____

____

____

letter scale

____

____

____

package scale

____

____

____

platform scale

____

____

____

bath scale

____

____

____

infant scale

____

____

____

spring scale

____

____

____

diet/food scale

____

____

____

Date Time

Reading Scales

Read each scale and record the weight.

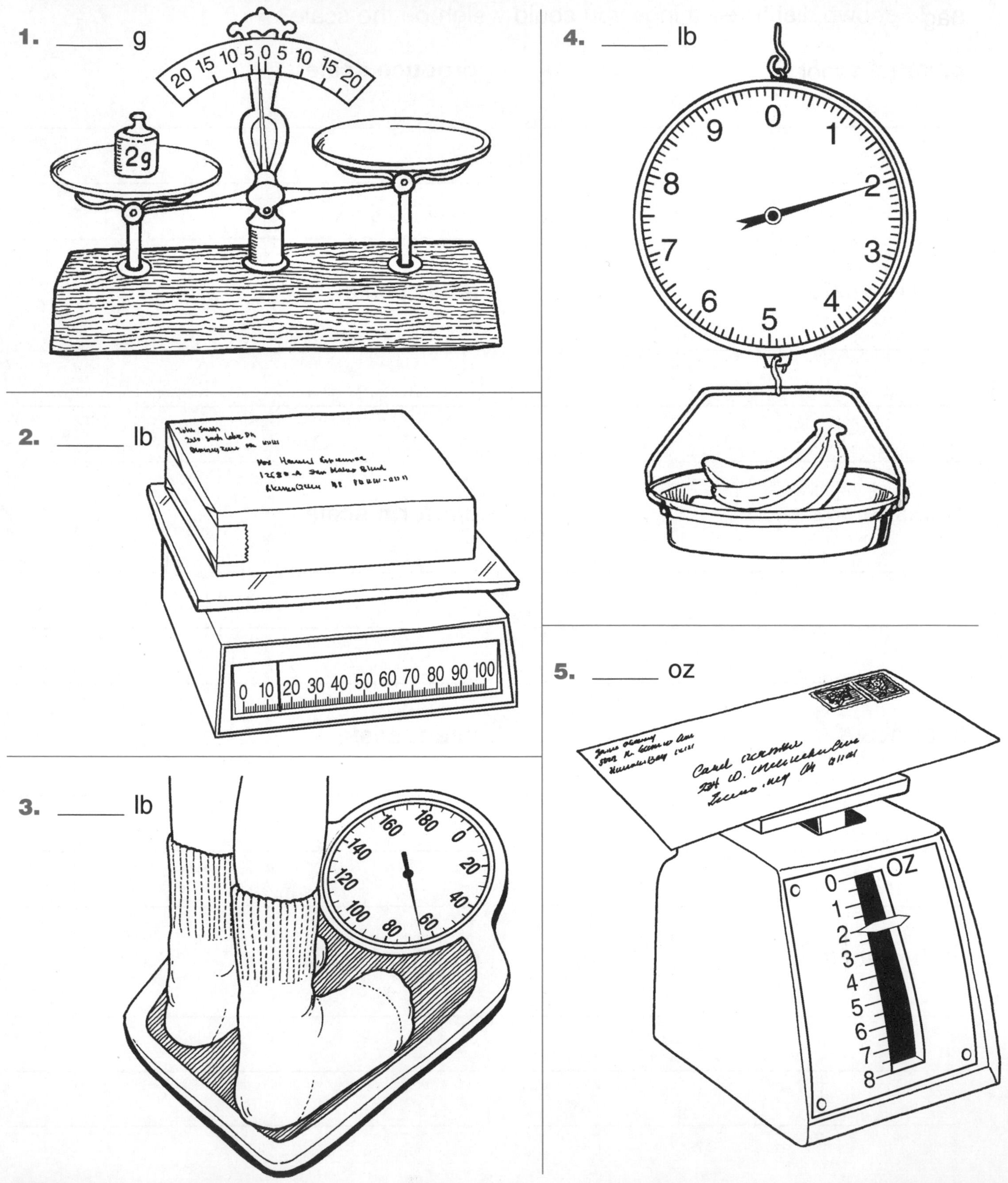

Date Time

Silly Stories

Refer to the Adult Weights of North American Animals Poster on journal pages 206 and 207 to solve the number stories.

1. If a 30-pound raccoon and a 150-pound deer both stood on a scale, what weight would the scale show?

   ______

2. If five 40-pound beavers climbed on one side of a pan balance, which animal might sit on the other pan so they balance?

   ______

   ______

3. If a 3,000-pound beluga whale, a 6,000-pound pilot whale, a 50,000-pound gray whale, and an 80,000-pound right whale lay on a platform scale (it would have to be huge!), what weight would the scale show? ______

   Which single whale could weigh this much? ______

4. One side of a pan balance has 50 three-pound Gila monsters. The other side of the pan balance has 10 five-pound snowshoe hares. How many of which animal could you add to one of the pans so that the pans balance? ______

   Would the animals go on the pan with the Gila monsters or the snowshoe hares? ______

5. Write and solve a problem of your own.

   ______

   ______

   ______

Date Time

Math Boxes 10.4

1. Complete the bar graph.

Eli biked 4 miles.

Kate biked 5 miles.

Joe biked 2 miles.

Miles			
5			
4			
3			
2			
1			
0	Eli	Kate	Joe

SRB 80 81

2. Fill in the oval next to the best estimate.

747 + 932 = _____

0 about 1,500

0 about 1,700

0 about 2,000

0 about 2,500

SRB 167

3. Write six numbers that are factors of 28.

_____ _____ _____

_____ _____ _____

SRB 37

4. Use your Pattern-Block Template to trace three shapes that are regular polygons.

SRB 95

5. Draw two ways to show $\frac{5}{4}$.

6. Draw a line segment $1\frac{3}{4}$ inches long.

Draw a line segment $\frac{1}{2}$ inch longer than the one you just drew.

SRB 125–127

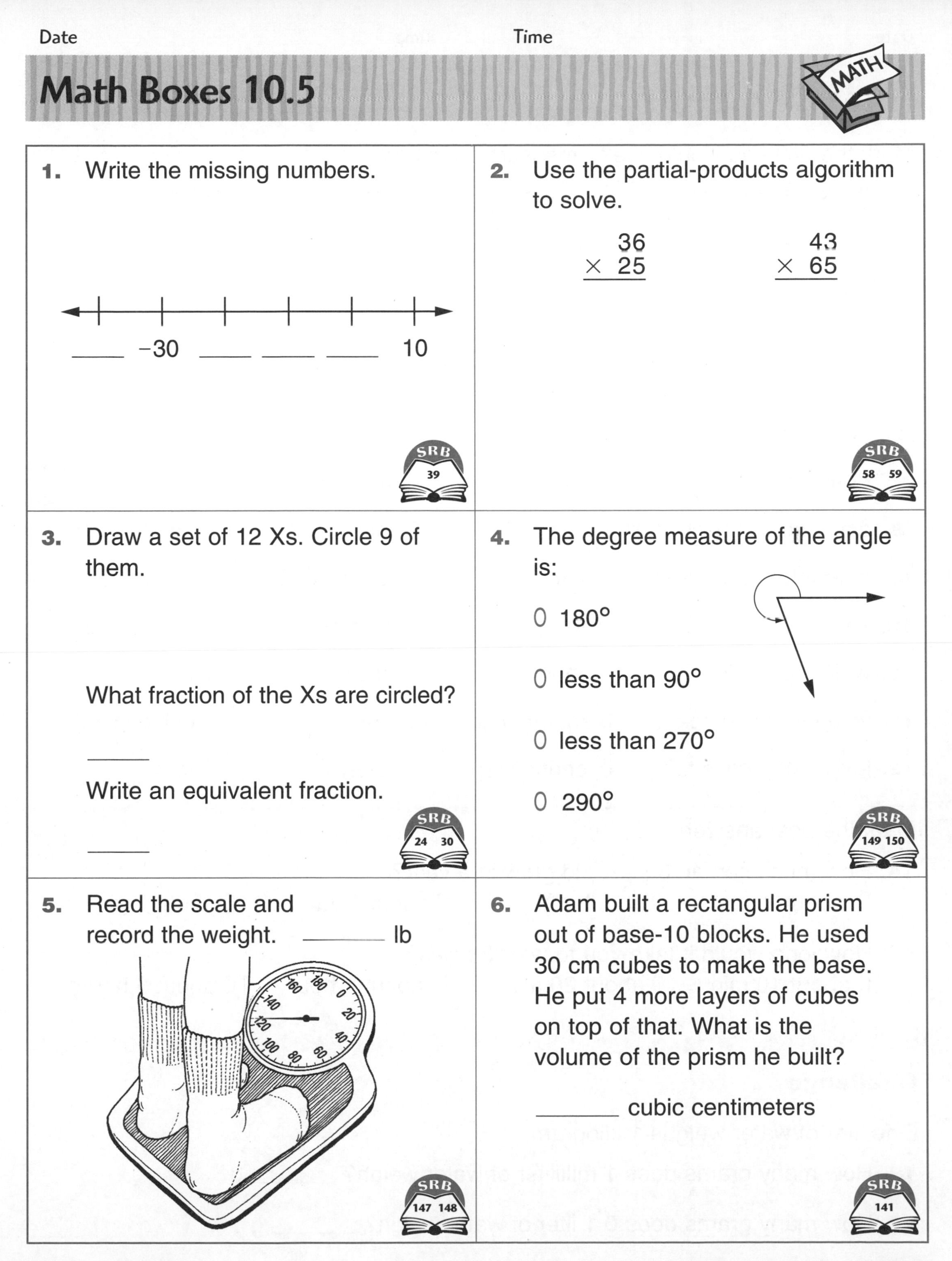

Math Boxes 10.5

MATH

1. Write the missing numbers.

____ −30 ____ ____ ____ 10

SRB 39

2. Use the partial-products algorithm to solve.

$$\begin{array}{r} 36 \\ \times\ 25 \\ \hline \end{array} \qquad \begin{array}{r} 43 \\ \times\ 65 \\ \hline \end{array}$$

SRB 58 59

3. Draw a set of 12 Xs. Circle 9 of them.

What fraction of the Xs are circled?

____

Write an equivalent fraction.

____

SRB 24 30

4. The degree measure of the angle is:

0 180°

0 less than 90°

0 less than 270°

0 290°

SRB 149 150

5. Read the scale and record the weight. ______ lb

SRB 147 148

6. Adam built a rectangular prism out of base-10 blocks. He used 30 cm cubes to make the base. He put 4 more layers of cubes on top of that. What is the volume of the prism he built?

______ cubic centimeters

SRB 141

Date Time

Units of Measure

Mark the unit you would use to measure each item.

1. thickness of a dime	0 millimeter	0 gram	0 foot
2. flour used in cooking	0 gallon	0 cup	0 liter
3. gasoline for a car	0 fluid ounce	0 ton	0 gallon
4. distance to the moon	0 foot	0 square mile	0 kilometer
5. area of a floor	0 square foot	0 cubic foot	0 foot
6. package of meat	0 yard	0 ounce	0 ton
7. draperies	0 kilometer	0 millimeter	0 yard
8. diameter of a basketball	0 mile	0 inch	0 square inch
9. perimeter of a garden	0 yard	0 square yard	0 centimeter
10. spices in a recipe	0 teaspoon	0 pound	0 fluid ounce
11. weight of a nickel	0 pound	0 gram	0 inch
12. volume of a suitcase	0 square inch	0 foot	0 cubic inch
13. length of a cat's tail	0 centimeter	0 meter	0 yard

Mark the best answer.

14. How much can an 8-year-old grow in a year?

0 about 2 in. 0 about 2 ft 0 about 1 yd 0 about 1 m

15. How long would it take you to walk 3 miles?

0 about 10 min 0 about 20 min 0 about 1 hour 0 about 5 hours

Challenge

One liter of water weighs 1 kilogram.

16. How many grams does 1 milliliter of water weigh? ______ g

17. How many grams does 0.1 liter of water weigh? ______ g

Date Time

Metric Weights

1. Two regular-size paper clips weigh about 1 gram.
   a. About how many paper clips would weigh 10 grams? __________
   b. About how many would weigh 1 kilogram? __________
   c. 0.5 kilogram? __________

2. One ounce is about 30 grams.
   a. About how many regular-size paper clips are in 1 ounce? __________
   b. 1 pound? __________

3. About how many grams does a box of 100 paper clips weigh if the empty box weighs about 5 grams? __________

4. A ream of paper has 500 sheets. Most reams of copying paper weigh a little more than 2 kilograms each. About how many grams does 1 sheet of paper weigh? __________

Review Solve.

5. $\begin{array}{r} 35 \\ \times\ 4 \\ \hline \end{array}$

6. $\begin{array}{r} 62 \\ \times\ 3 \\ \hline \end{array}$

7. $\begin{array}{r} 285 \\ \times\ 6 \\ \hline \end{array}$

8. Write a number story for one of the problems you just solved.

____

____

____

____

Date Time

Math Boxes 10.6

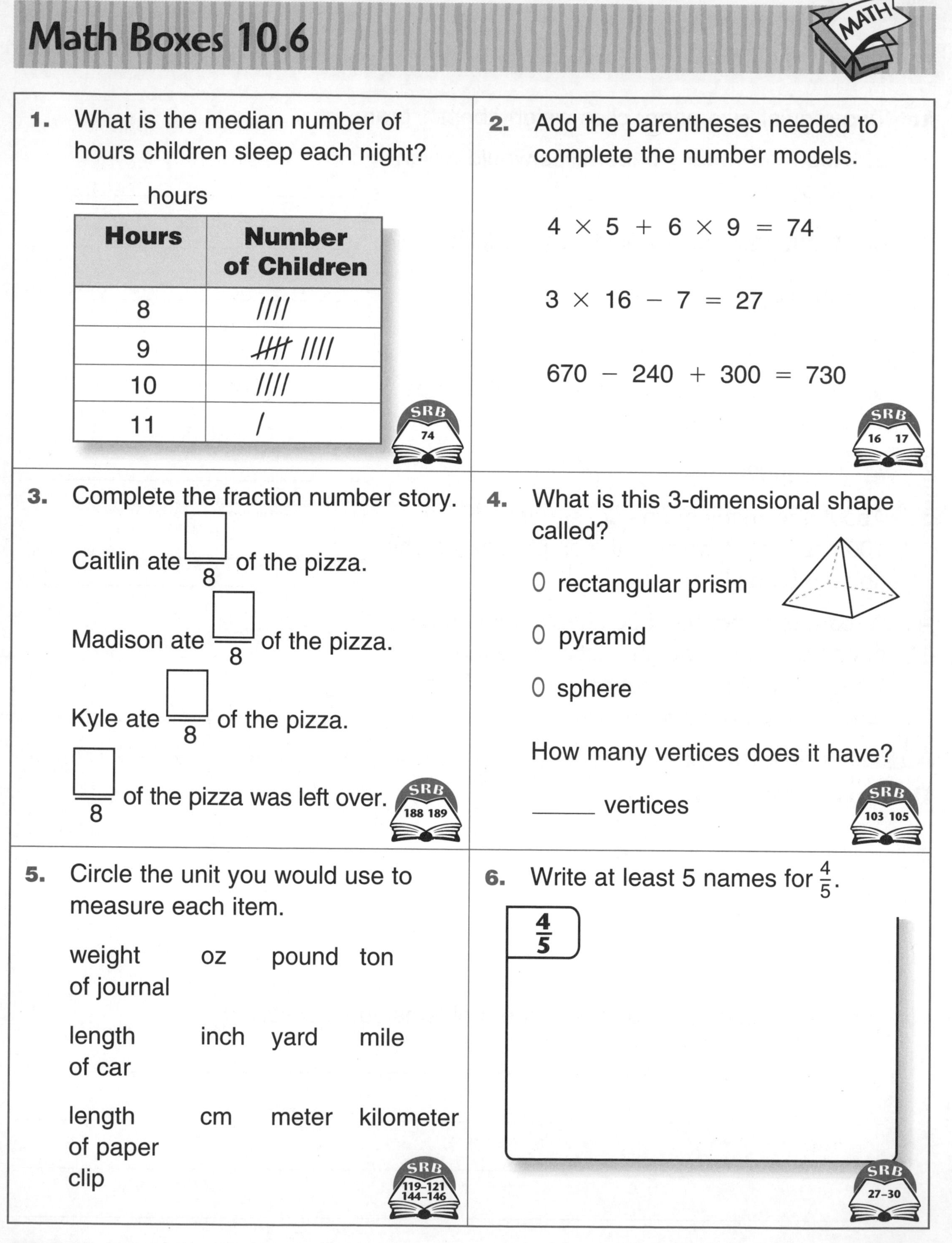

1. What is the median number of hours children sleep each night?

_____ hours

Hours	Number of Children
8	////
9	~~////~~ ////
10	////
11	/

SRB 74

2. Add the parentheses needed to complete the number models.

$4 \times 5 + 6 \times 9 = 74$

$3 \times 16 - 7 = 27$

$670 - 240 + 300 = 730$

SRB 16 17

3. Complete the fraction number story.

Caitlin ate $\frac{\square}{8}$ of the pizza.

Madison ate $\frac{\square}{8}$ of the pizza.

Kyle ate $\frac{\square}{8}$ of the pizza.

$\frac{\square}{8}$ of the pizza was left over.

SRB 188 189

4. What is this 3-dimensional shape called?

0 rectangular prism

0 pyramid

0 sphere

How many vertices does it have?

_____ vertices

SRB 103 105

5. Circle the unit you would use to measure each item.

weight of journal	oz	pound	ton
length of car	inch	yard	mile
length of paper clip	cm	meter	kilometer

SRB 119–121 144–146

6. Write at least 5 names for $\frac{4}{5}$.

$\frac{4}{5}$

SRB 27–30

Date Time

A Mean (or Average) Number of Children

Activity 1 Make a bar graph of the data in the table.

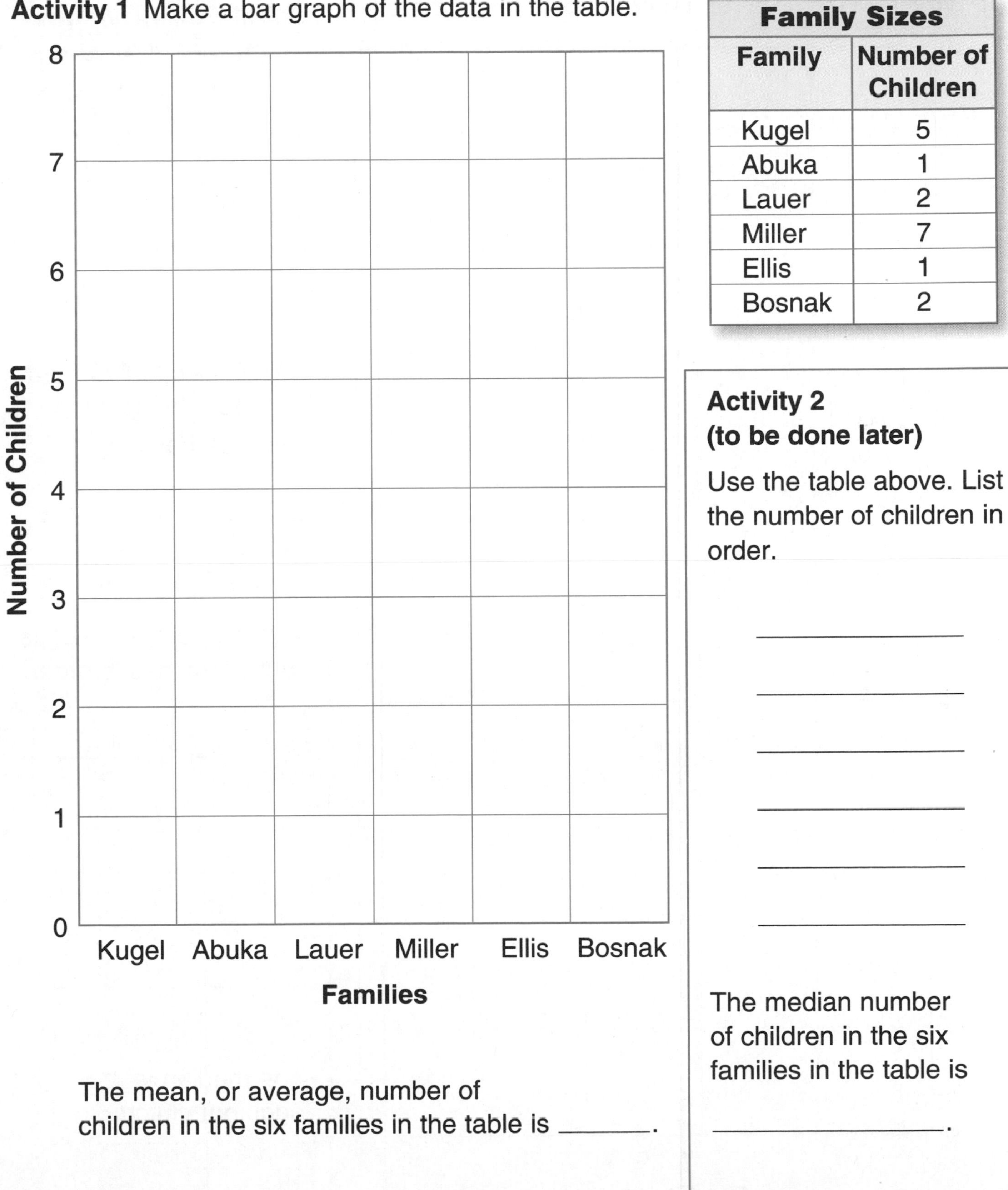

Family Sizes	
Family	**Number of Children**
Kugel	5
Abuka	1
Lauer	2
Miller	7
Ellis	1
Bosnak	2

The mean, or average, number of children in the six families in the table is ______.

Activity 2 (to be done later)

Use the table above. List the number of children in order.

______

______

______

______

______

______

The median number of children in the six families in the table is

______.

Date Time

A Mean (or Average) Number of Eggs

Activity 1 Make a bar graph of the data in the table.

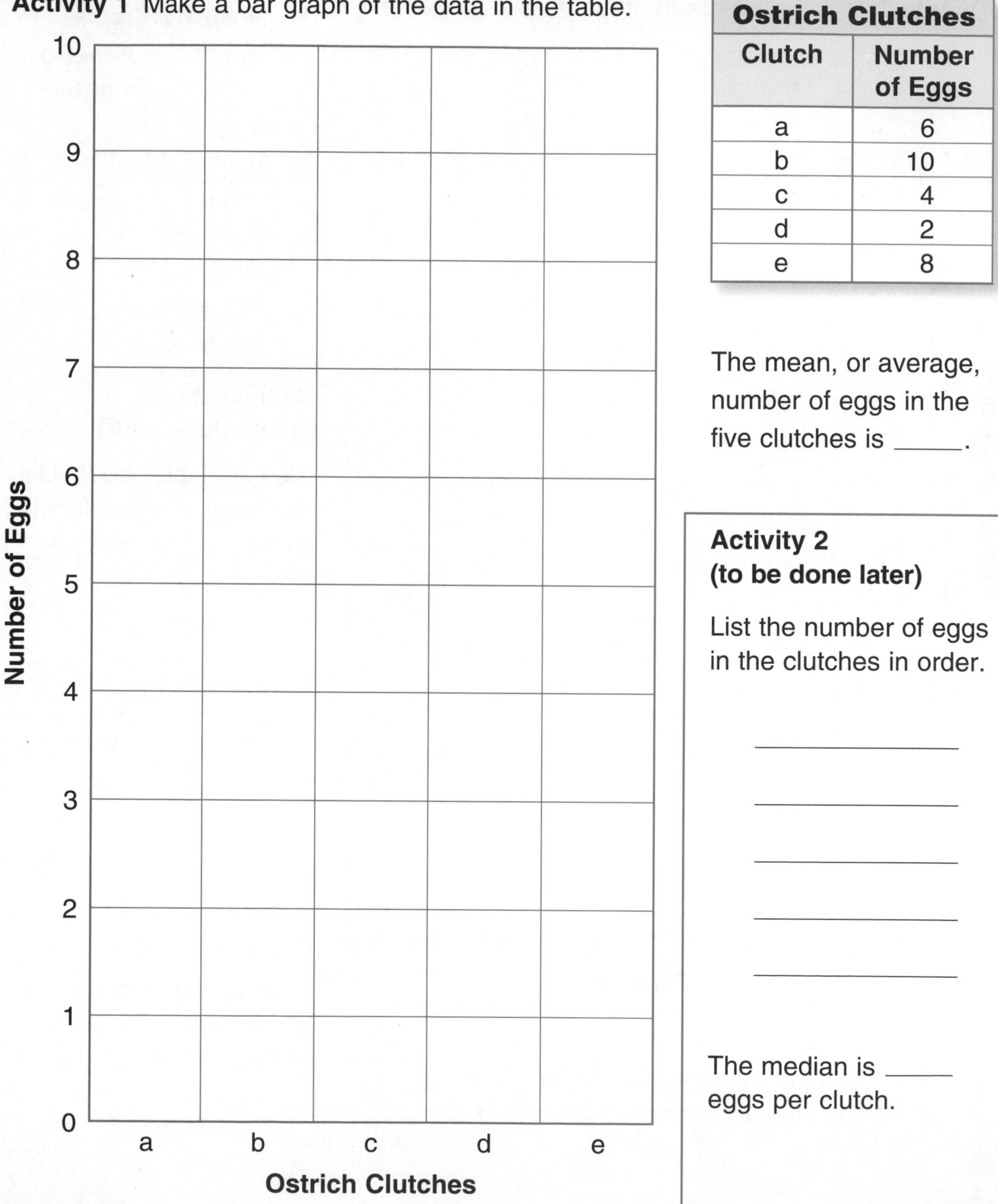

Ostrich Clutches	
Clutch	**Number of Eggs**
a	6
b	10
c	4
d	2
e	8

The mean, or average, number of eggs in the five clutches is _____.

Activity 2
(to be done later)

List the number of eggs in the clutches in order.

____

____

____

____

____

The median is _____ eggs per clutch.

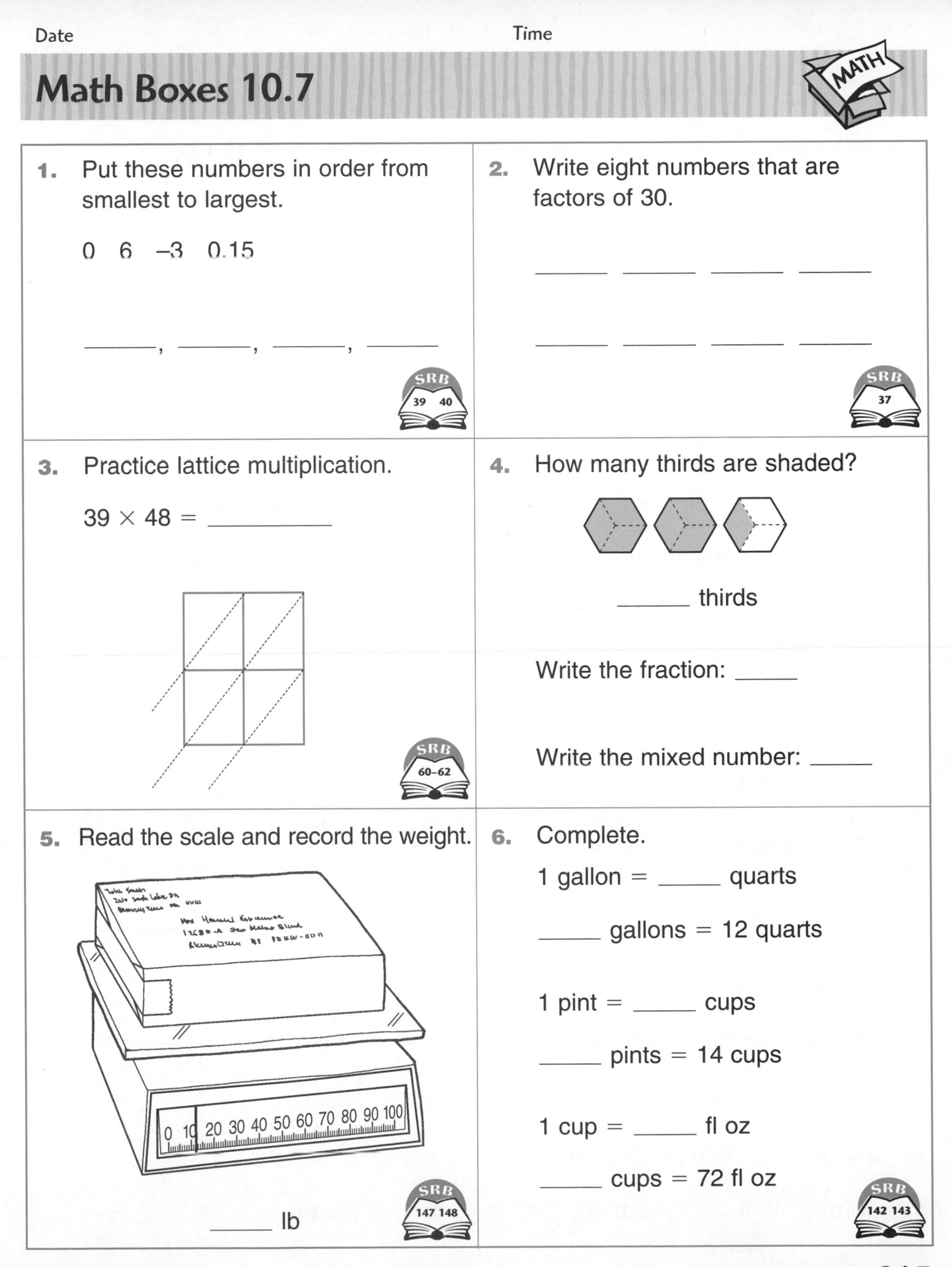

Date Time

Math Boxes 10.7

MATH

1. Put these numbers in order from smallest to largest.

0 6 –3 0.15

______, ______, ______, ______

SRB 39 40

2. Write eight numbers that are factors of 30.

______ ______ ______ ______

______ ______ ______ ______

SRB 37

3. Practice lattice multiplication.

$39 \times 48 =$ __________

SRB 60–62

4. How many thirds are shaded?

______ thirds

Write the fraction: ______

Write the mixed number: ______

5. Read the scale and record the weight.

______ lb

SRB 147 148

6. Complete.

1 gallon = ______ quarts

______ gallons = 12 quarts

1 pint = ______ cups

______ pints = 14 cups

1 cup = ______ fl oz

______ cups = 72 fl oz

SRB 142 143

Date Time

Finding the Median and the Mean

1. The median (middle) arm span in my class is about _____ inches.

2. The mean (average) arm span in my class is about _____ inches.

3. Look at page 253 in your journal. Use the measurements for an adult and the *second* measurements for yourself to find the median and mean arm spans and heights for your group. Record the results in the table below.

   a. Find the median and mean arm spans of the *adults* for your group.

   b. Find the median and mean arm spans of the *children* for your group.

   c. Find the median and mean heights of the *adults* for your group.

   d. Find the median and mean heights of the *children* for your group.

Summary of Measurements for Your Group

Measure	Adults	Children
Median arm span		
Mean arm span		
Median height		
Mean height		

Find the mean of each set of data.

4. High temperatures: 56°F, 62°F, 74°F, 68°F Mean: __________ °F

5. Low temperatures: 32°F, 42°F, 58°F, 60°F Mean: __________ °F

6. Ticket sales: $710, $650, $905 Mean: $ __________

7. Throws: 40 ft, 32 ft, 55 ft, 37 ft, 43 ft, 48 ft Mean: __________ ft

Use with Lesson 10.8.

Date Time

Measurement Number Stories

1. The gas tank of Mrs. Rone's car holds about 12 gallons. About how many gallons are in the tank when the gas gauge shows the tank to be $\frac{3}{4}$ full?

workspace

____

2. When the gas tank of Mrs. Rone's car is about half empty, she stops to fill the tank. If gas costs $1.25 per gallon, about how much does it cost to fill the tank?

____

Harry's room measures 11 feet by 13 feet. The door to his room is 3 feet wide. He wants to put a wooden border, or baseboard, around the base of the walls.

3. Draw a diagram of Harry's room on the grid below. Show where the door is. Let each side of a grid square equal 1 foot.

4. How many feet of baseboard must Harry buy? __________

5. How many yards is that? __________

6. If baseboard costs $4.00 a yard, how much will Harry pay? __________

workspace

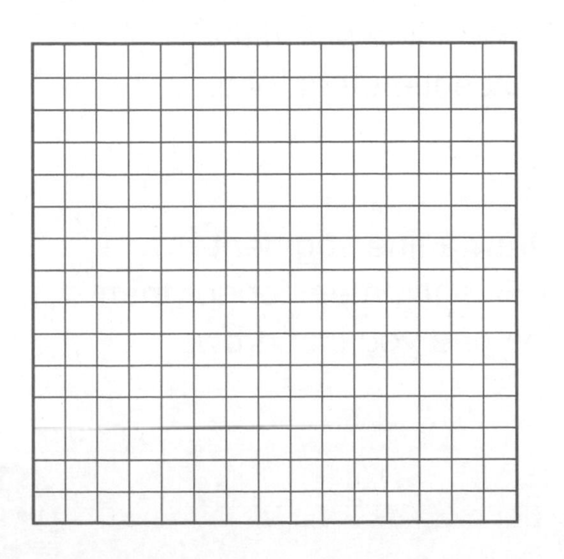

Date Time

Math Boxes 10.8

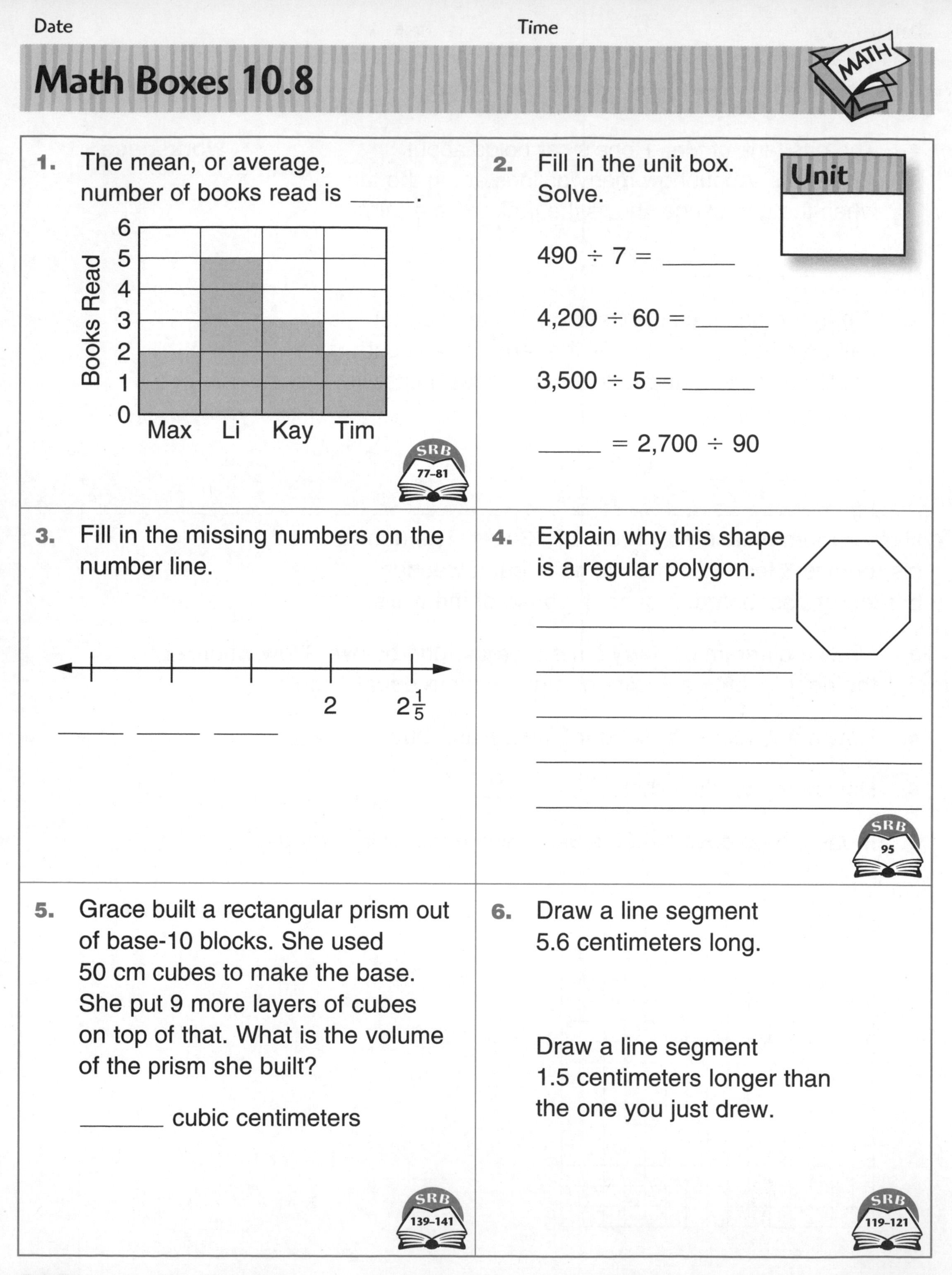

1. The mean, or average, number of books read is _____.

2. Fill in the unit box. Solve.

Unit

490 ÷ 7 = _____

4,200 ÷ 60 = _____

3,500 ÷ 5 = _____

_____ = 2,700 ÷ 90

3. Fill in the missing numbers on the number line.

_____ _____ _____

4. Explain why this shape is a regular polygon.

____

____

____

____

____

5. Grace built a rectangular prism out of base-10 blocks. She used 50 cm cubes to make the base. She put 9 more layers of cubes on top of that. What is the volume of the prism she built?

_______ cubic centimeters

6. Draw a line segment 5.6 centimeters long.

Draw a line segment 1.5 centimeters longer than the one you just drew.

Date Time

Calculator Memory

For each problem:

- Press the keys on the calculator.
- Guess what number is in memory.
- Record your guess.
- Press (MRC) to check your guess.
- Record the answer.

After each problem, press (MRC) twice and (ON/C) to clear everything.
The display should look like this [0.] before you start a new problem.

	Press these keys	Your Guess	Press (MRC) Answer
1.	(MRC) (ON/C) (7) (M+) (9) (M+)	______	______
2.	(2) (0) (M+) (1) (6) (M−)	______	______
3.	(5) (M+) (1) (0) (M+) (2) (0) (M+)	______	______
4.	(1) (2) (+) (8) (M+) (6) (M−)	______	______
5.	(1) (5) (−) (9) (M+) (2) (M−)	______	______
6.	(2) (5) (M+) (7) (+) (8) (M−)	______	______
7.	(3) (0) (M+) (1) (5) (−) (5) (M+)	______	______
8.	(2) (+) (2) (M+) (2) (−) (2) (M+) (2) (+) (2) (M+)	______	______

Review Solve.

9. $8 + (2 \times 7) =$ _____

10. $(8 + 2) \times 7 =$ _____

11. $(24 - 9) \times 2 =$ _____

12. $24 - (9 \times 2) =$ _____

13. _____ $= (36 - 22) - 14$

14. _____ $= 36 - (22 - 14)$

15. _____ $= 35 + (20 - 15)$

16. _____ $= (35 + 20) - 15$

17. _____ $= 35 - (20 + 15)$

18. _____ $= (35 - 20) + 15$

Date Time

Math Boxes 10.9

1. Complete.

Area of Base (square cm)	Height (cm)	Volume (cubic cm)
40	7	
35	6	
100	30	
40	50	

SRB 141

2. Write 5 fractions greater than $\frac{4}{10}$.

____ ____ ____ ____ ____

Write 5 fractions less than $\frac{4}{10}$.

____ ____ ____ ____ ____

Write 3 other names for $\frac{4}{10}$.

____ ____ ____

SRB 31 32

3. Shade $\frac{3}{5}$ of the rectangle.

What fraction is not shaded?

____

SRB 22 23

4. Draw an angle that measures approximately 90°.

An angle that measures 90° is called a __________ angle.

SRB 90

5. Name 4 objects that weigh less than 1 pound.

____

____

____

____

SRB 144 145

6. Complete.

1 quart = ______ pints

______ quarts = 16 pints

1 quart = ______ fl oz

______ quarts = 96 fl oz

1 gallon = ______ fl oz

SRB 142 143

Date Time

Math Boxes 10.10

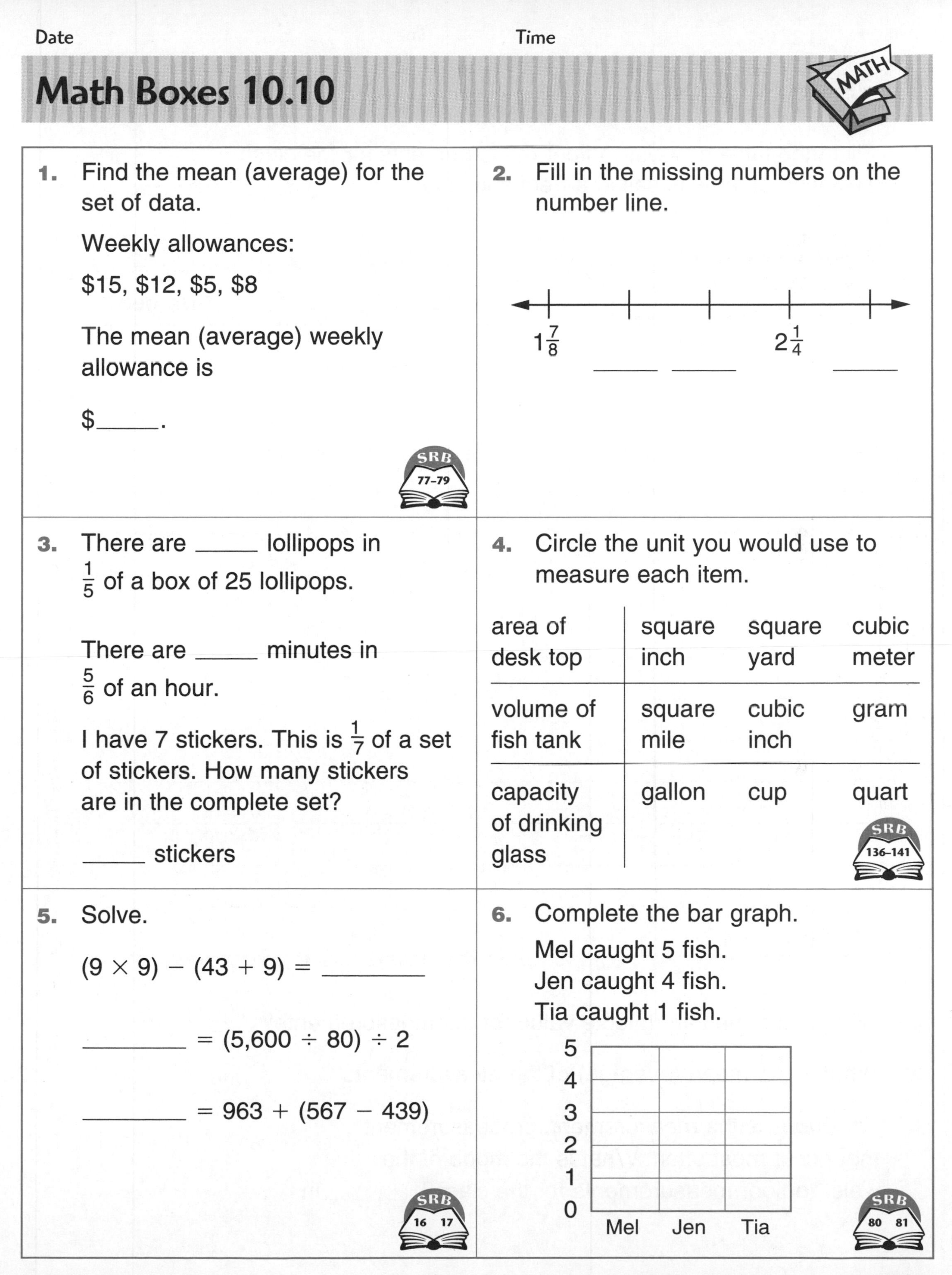

1. Find the mean (average) for the set of data.

Weekly allowances:

\$15, \$12, \$5, \$8

The mean (average) weekly allowance is

\$_____.

SRB 77–79

2. Fill in the missing numbers on the number line.

$1\frac{7}{8}$ _____ _____ $2\frac{1}{4}$ _____

3. There are _____ lollipops in $\frac{1}{5}$ of a box of 25 lollipops.

There are _____ minutes in $\frac{5}{6}$ of an hour.

I have 7 stickers. This is $\frac{1}{7}$ of a set of stickers. How many stickers are in the complete set?

_____ stickers

4. Circle the unit you would use to measure each item.

area of desk top	square inch	square yard	cubic meter
volume of fish tank	square mile	cubic inch	gram
capacity of drinking glass	gallon	cup	quart

SRB 136–141

5. Solve.

$(9 \times 9) - (43 + 9) =$ _______

_______ $= (5{,}600 \div 80) \div 2$

_______ $= 963 + (567 - 439)$

SRB 16 17

6. Complete the bar graph.

Mel caught 5 fish.
Jen caught 4 fish.
Tia caught 1 fish.

SRB 80 81

Date Time

Frequency Table

1. Fill in the table of waist-to-floor measurements for the class. This kind of table is called a frequency table.

Waist-to-Floor Measurement (inches)	Frequency	
	Tallies	Number
Total =		

2. What is the median (middle value) of the measurements? _______ in.

3. What is the mean (average) of the measurements? _______ in.

4. The *mode* is the measurement, or measurements, that occur most often. What is the mode of the waist-to-floor measurements for the class? _______ in.

Use with Lesson 10.10.

Date Time

Bar Graph

Make a bar graph of the data in the frequency table on journal page 272.

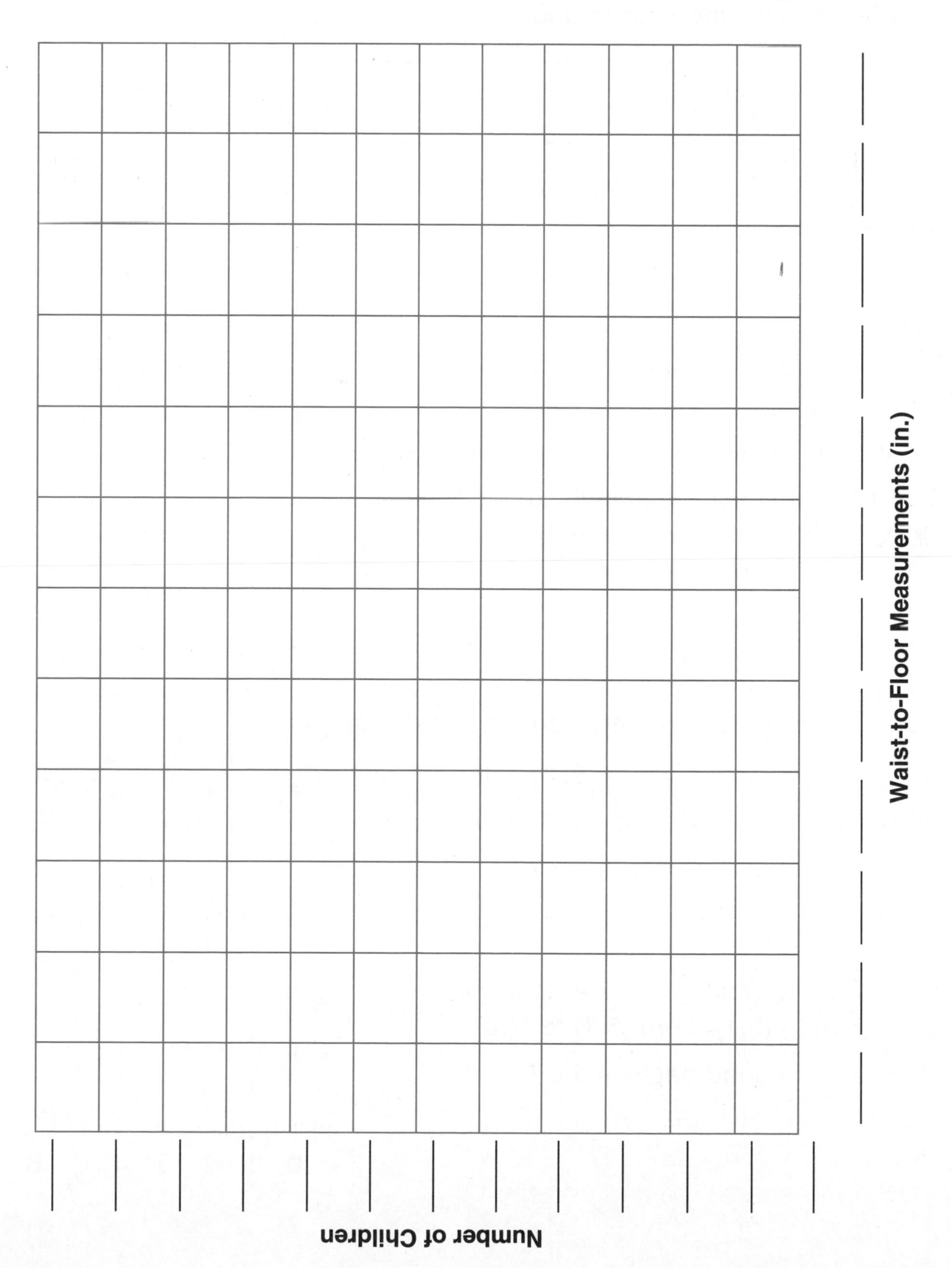

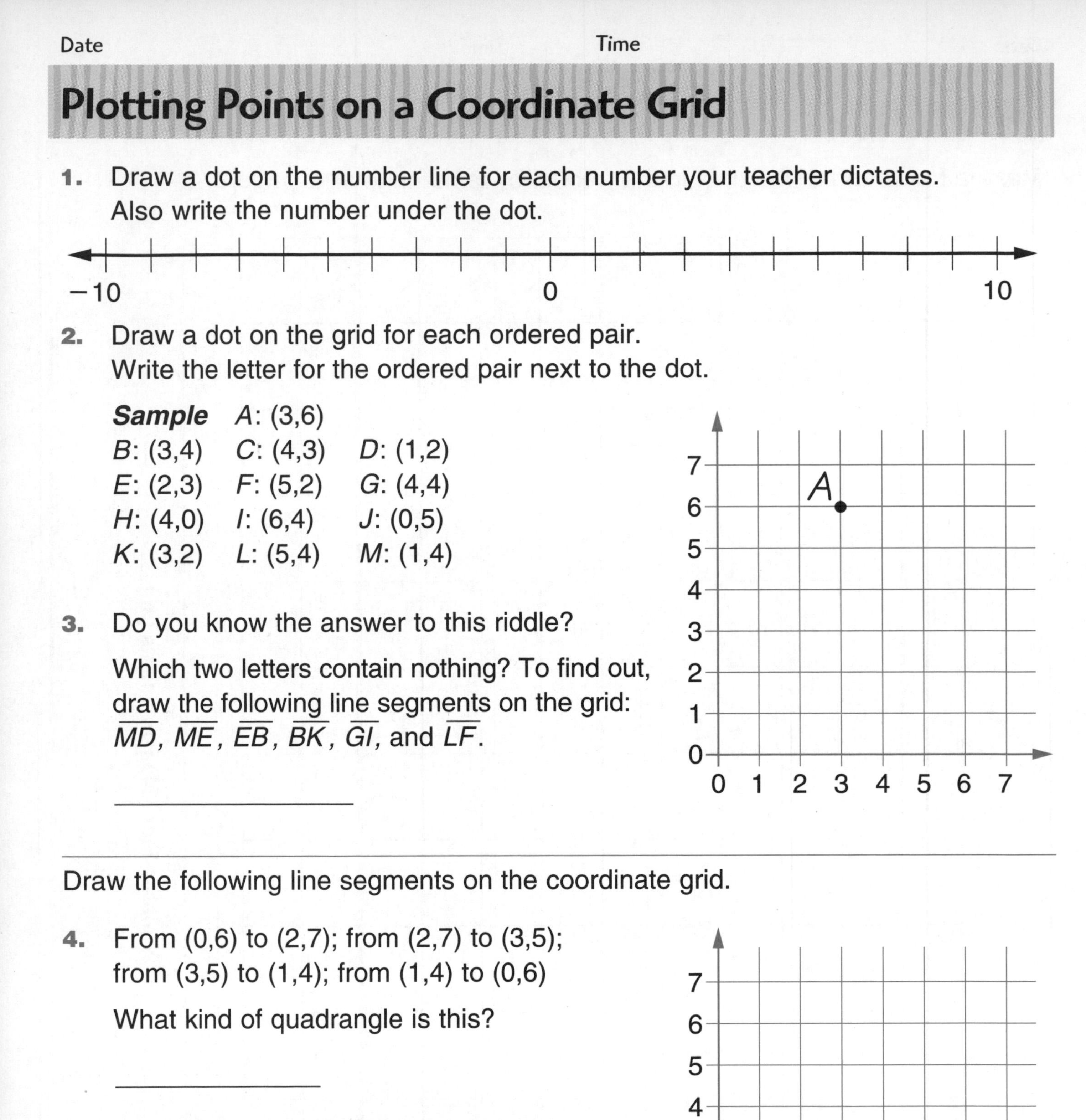

Date Time

Plotting Points on a Coordinate Grid

1. Draw a dot on the number line for each number your teacher dictates.
Also write the number under the dot.

2. Draw a dot on the grid for each ordered pair.
Write the letter for the ordered pair next to the dot.

Sample *A*: (3,6)

B: (3,4) *C*: (4,3) *D*: (1,2)
E: (2,3) *F*: (5,2) *G*: (4,4)
H: (4,0) *I*: (6,4) *J*: (0,5)
K: (3,2) *L*: (5,4) *M*: (1,4)

3. Do you know the answer to this riddle?

Which two letters contain nothing? To find out, draw the following line segments on the grid: $\overline{MD}$, $\overline{ME}$, $\overline{EB}$, $\overline{BK}$, $\overline{GI}$, and $\overline{LF}$.

____

Draw the following line segments on the coordinate grid.

4. From (0,6) to (2,7); from (2,7) to (3,5); from (3,5) to (1,4); from (1,4) to (0,6)

What kind of quadrangle is this?

____

5. From (7,0) to (7,4); from (7,4) to (5,3); from (5,3) to (5,1); from (5,1) to (7,0)

What kind of quadrangle is this?

____

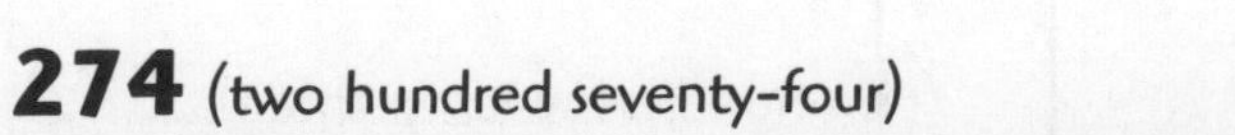

Date Time

Using a Commuter Railroad Timetable

Solve the problems. Use your tool-kit clock if you need help.

Train Schedule	
South Chicago	11:46 A.M.
83rd Street	11:49
Cheltenham	11:51
South Shore	11:55
Bryn Mawr	11:57
59th Street	12:04 P.M.
Hyde Park	12:08
Kenwood	12:09
McCormick Place	12:14
18th Street	12:15
Van Buren Street	12:19
Randolph Street	12:22

1. About how many minutes is the trip from South Chicago to Randolph Street?

   ____

2. At what station will the train stop about 22 minutes after it leaves South Chicago?

   ____

3. At what station did the train stop about 25 minutes before it got to Randolph Street?

   ____

4. At what station does the train stop halfway through the trip—when about half of the total trip time has passed?

   ____

5. Marci got on the train at Cheltenham and got off at Kenwood. About how long was she on the train? ______________

6. Make up two problems. Ask your partner to solve them.

____

____

____

____

____

____

Math Boxes 10.11

1. Find the distance between each pair of numbers.

4 and −19 ________

−23 and 46 ________

1,000 and −7,000 ________

2. Points scored by players in a basketball game:

15, 22, 11, 12, 5

The mean (average) number of points is _____.

SRB 77–79

3. What is the mode of the test scores for the class? ______%

Test Score	Number of Children
100%	///
95%	~~////~~
90%	~~////~~ ///
85%	////

SRB 75 76

4. Complete.

Area of Base (square cm)	Height (cm)	Volume (cubic cm)
60	6	
45	4	
200	70	
80	80	

SRB 141

5. Read the scale and record the weight.

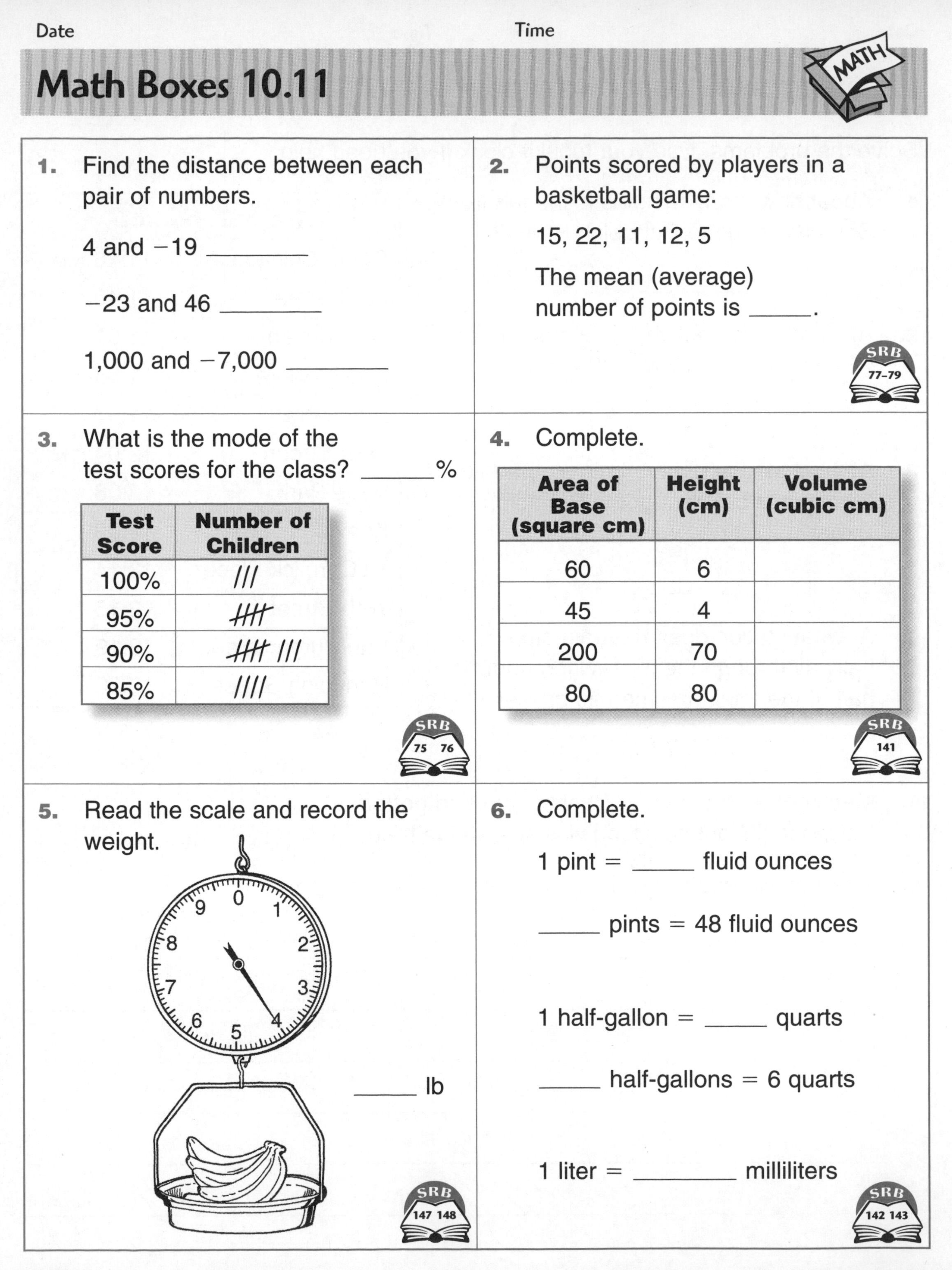

_____ lb

SRB 147 148

6. Complete.

1 pint = _____ fluid ounces

_____ pints = 48 fluid ounces

1 half-gallon = _____ quarts

_____ half-gallons = 6 quarts

1 liter = ________ milliliters

SRB 142 143

Date Time

Math Boxes 10.12

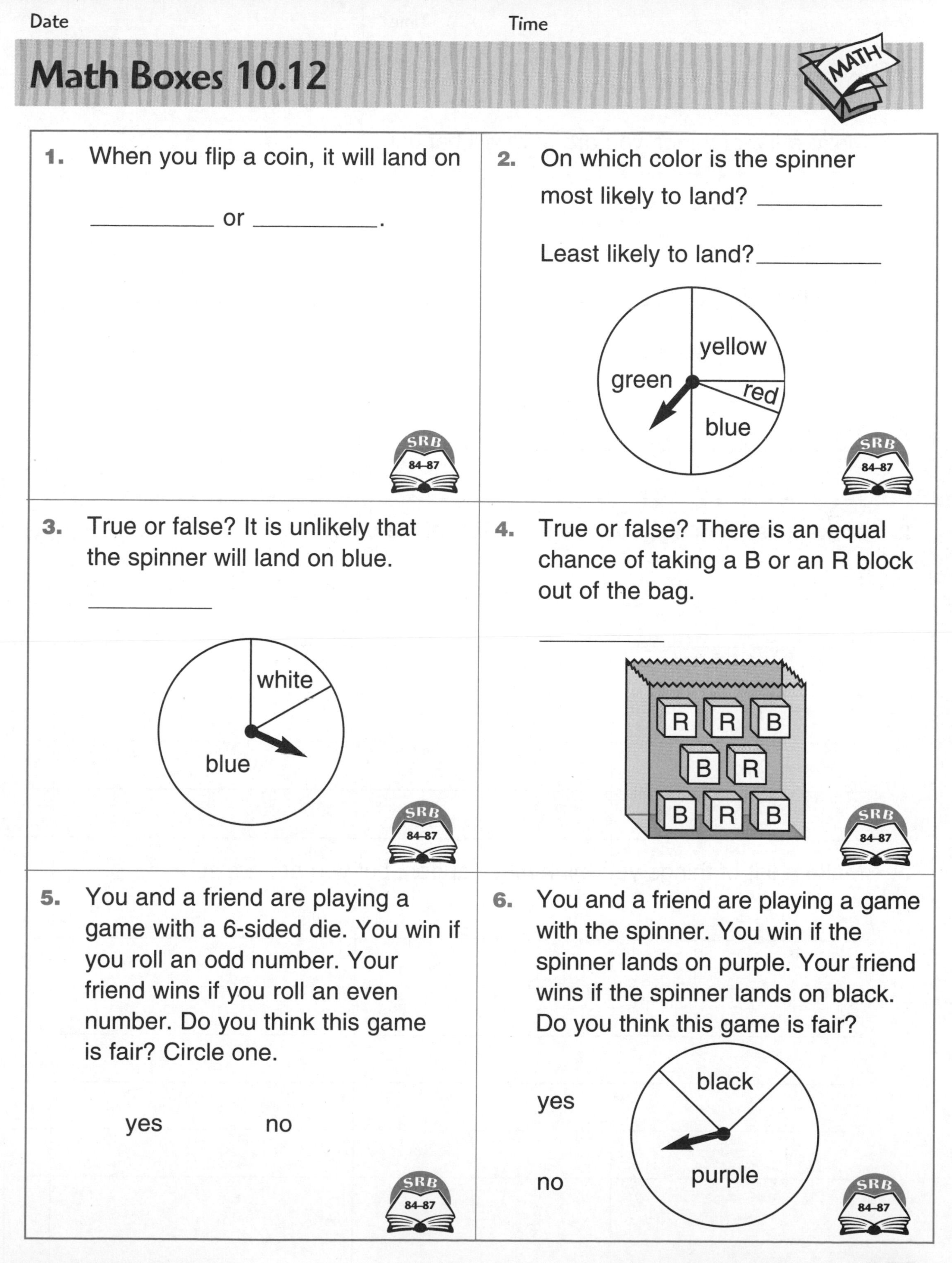

1. When you flip a coin, it will land on

_________ or _________.

SRB 84–87

2. On which color is the spinner most likely to land? _________

Least likely to land? _________

SRB 84–87

3. True or false? It is unlikely that the spinner will land on blue.

____

SRB 84–87

4. True or false? There is an equal chance of taking a B or an R block out of the bag.

____

SRB 84–87

5. You and a friend are playing a game with a 6-sided die. You win if you roll an odd number. Your friend wins if you roll an even number. Do you think this game is fair? Circle one.

yes no

SRB 84–87

6. You and a friend are playing a game with the spinner. You win if the spinner lands on purple. Your friend wins if the spinner lands on black. Do you think this game is fair?

yes

no

SRB 84–87

Date Time

Can You Be Sure?

1. Make a list of things you are *sure* will happen.

2. Make a list of things you are sure will *not* happen.

3. Make a list of things you think *may* happen, but you are not sure.

Date Time

Reading and Writing Numbers

Write the value of 7 for each column below.

L	K	J	I	H	G	F	E	D	.	C	B	A
hundred-millions	ten-millions	millions	hundred-thousands	ten-thousands	thousands	hundreds	tens	ones	.	tenths	hundredths	thousandths
7	7	7	7	7	7	7	7	7	.	7	7	7

Example Column **K**: 70,000,000 or 70 million

1. Column **A**: ______
2. Column **G**: ______
3. Column **F**: ______
4. Column **I**: ______
5. Column **C**: ______
6. Column **B**: ______
7. Column **L**: ______

Write the numbers that your teacher dictates.

8. ______ 9. ______ 10. ______

11. ______ 12. ______ 13. ______

Date Time

Math Boxes 11.1

1. Plot and label each of the points listed below.

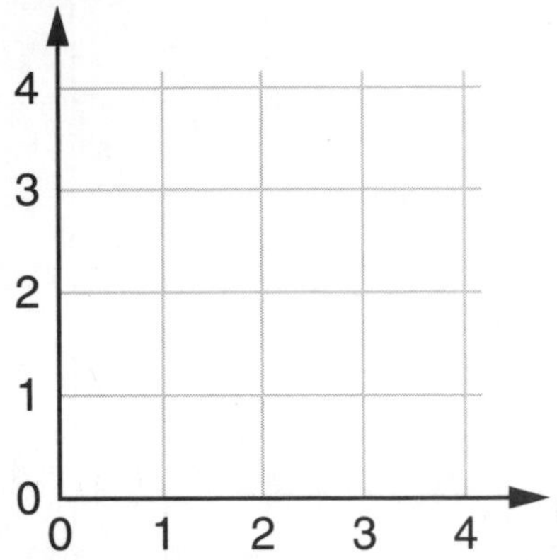

A: (1,4)

B: (2,2)

C: (3,1)

D: (4,3)

SRB 162 163

2. Cross out the names that do not belong.

$\frac{1}{5}$

$\frac{2}{10}$ $\frac{100}{500}$ $\frac{1}{8}$

$\frac{4}{5}$ $1\frac{1}{5}$ $\frac{5}{25}$

SRB 30

3. Weight in pounds of newborn babies: 11, 8, 8, 7, 6

The average (mean) weight is

_____ pounds.

The weight that occurs most often (mode) is _____ pounds.

SRB 75–79

4. Insert $<$, $>$, or $=$.

5×9 _____ 7×7

8×9 _____ 6×8

4×8 _____ 3×9

7×8 _____ 9×9

SRB 13 46 47

5. Use the partial-products algorithm to solve.

$$\begin{array}{r} 83 \\ \times\ 44 \\ \hline \end{array} \qquad \begin{array}{r} 72 \\ \times\ 36 \\ \hline \end{array}$$

SRB 58 59

6. There are 347 candles. A box holds 50 candles. How many full boxes of candles is that?

_____ boxes

How many candles are left over?

_____ candles

SRB 68

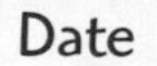

Date Time

Number Lines

Fill in the missing numbers.

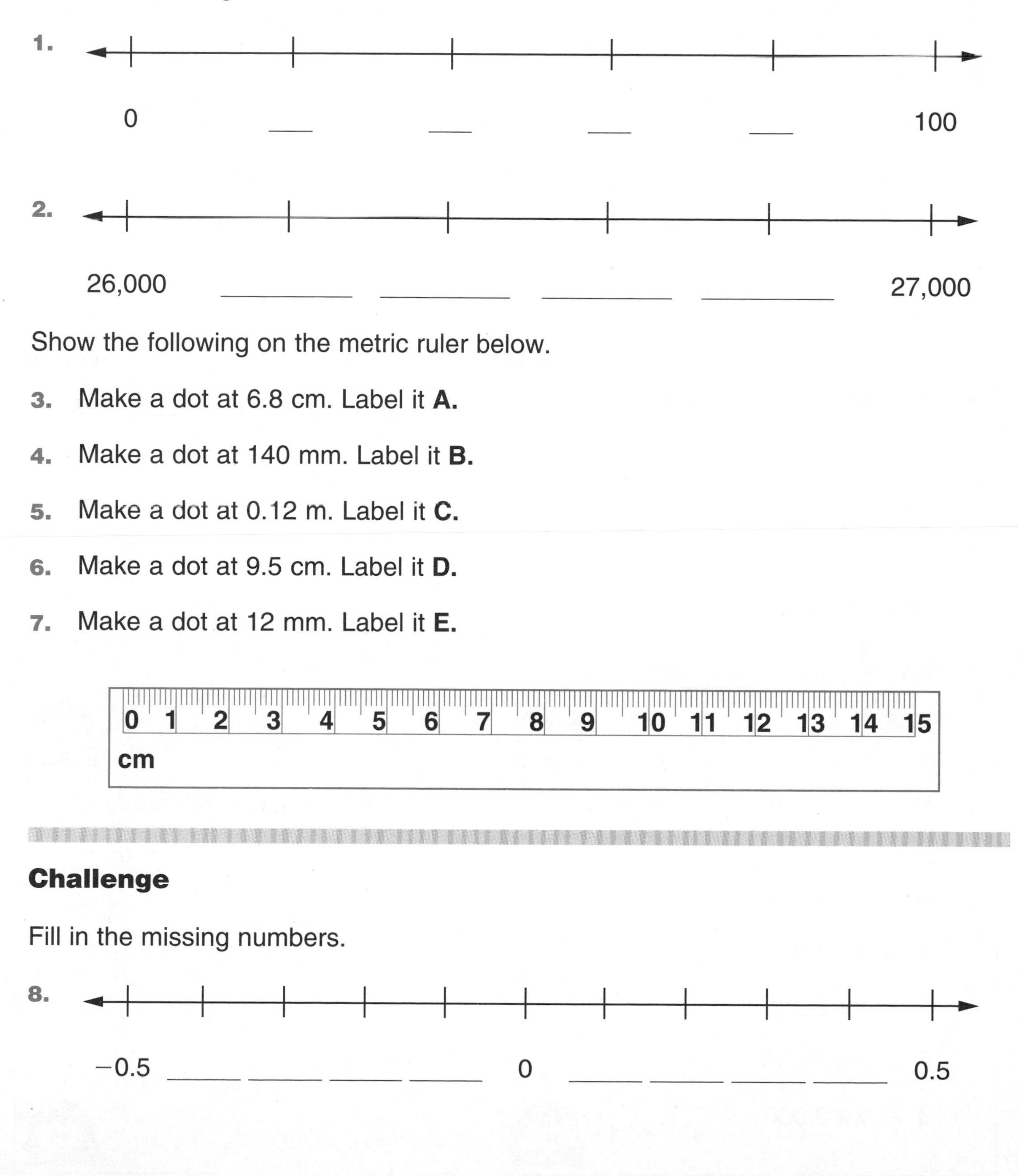

Show the following on the metric ruler below.

3. Make a dot at 6.8 cm. Label it **A.**

4. Make a dot at 140 mm. Label it **B.**

5. Make a dot at 0.12 m. Label it **C.**

6. Make a dot at 9.5 cm. Label it **D.**

7. Make a dot at 12 mm. Label it **E.**

Challenge

Fill in the missing numbers.

Date Time

Math Boxes 11.2

1. Complete.

Area of Base (square feet)	Height (feet)	Volume (cubic feet)
40	90	
20	70	
800	9	
50	80	

SRB 139–141

2. Complete.

2 gallons = _____ quarts

_____ gallons = 16 quarts

2 pints = _____ cups

_____ pints = 20 cups

2 cups = _____ fl oz

_____ cups = 32 fl oz

SRB 142–143

3. Circle the event(s) that you are sure will happen.

It will be sunny tomorrow.

A tossed quarter will land on either heads or tails.

A rolled die will land on 6.

SRB 84–86

4. Name 3 objects that weigh about 1 gram.

____

____

____

____

SRB 144 145

5. Fill in the oval next to the best estimate.

5,634 − 2,987 = ______

0 about 2,000

0 about 2,300

0 about 2,600

0 about 3,000

SRB 168

6. Write six numbers that are factors of 32.

_____ _____ _____

_____ _____ _____

SRB 37

Date Time

Coin-Toss Experiment

Work with a partner. You need 10 coins.

1. Each of you takes turns tossing the 10 coins.

   For each toss you make, record the number of heads and the number of tails in the table.

   Toss the coins 5 times in all.

   Then find the total number of heads and tails.

My Toss Record

Toss (10 coins)	Heads	Tails
1		
2		
3		
4		
5		
Total		

My total: heads ________ tails ________

My partner's total: heads ________ tails ________

Our partnership total: heads ________ tails ________

2. Record the number of heads and the number of tails for the whole class.

   Number of heads: ________ Number of tails: ________

3. Suppose a jar contains 1,000 pennies. The jar is turned over. The pennies are dumped onto a table and spread out. Write your best guess for the number of heads and tails.

   Number of heads: ________ Number of tails: ________

Date Time

Fractions

1. The shaded square is ONE. Write a name for the shaded part of each of the other shapes. The first one is done for you.

1

$\frac{1}{2}$ _____ _____ _____ _____ _____ _____

2. The shaded rectangle is ONE. Write a name for the shaded part of each of the other shapes.

1

_____ _____ _____

Write <, >, or = to compare fractions.

3. $\frac{3}{5}$ ☐ $\frac{1}{2}$ **4.** $\frac{3}{8}$ ☐ $\frac{3}{4}$ **5.** $\frac{4}{6}$ ☐ $\frac{5}{9}$ **6.** $\frac{3}{4}$ ☐ $\frac{9}{12}$ **7.** $\frac{7}{10}$ ☐ $\frac{6}{12}$

8. Choose one comparison above and explain how you found your answer.

____

____

9. Write numbers in this name-collection box.

$\frac{1}{4}$

10. Make your own name-collection box. Write a fraction on the tag. Write numbers in the box.

Date Time

Math Boxes 11.3

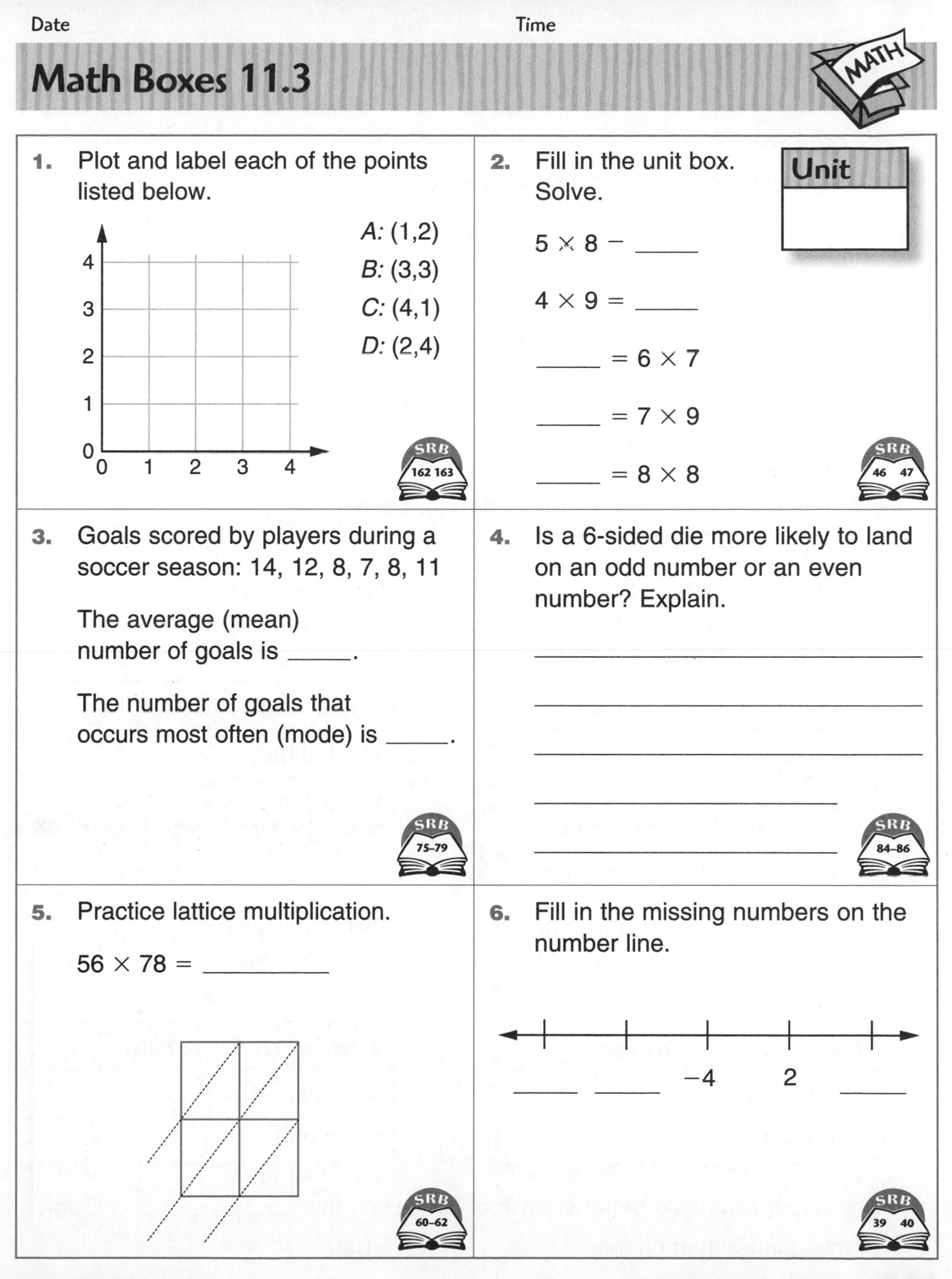

1. Plot and label each of the points listed below.

A: (1,2)
B: (3,3)
C: (4,1)
D: (2,4)

2. Fill in the unit box. Solve.

Unit

$5 \times 8 =$ _____

$4 \times 9 =$ _____

_____ $= 6 \times 7$

_____ $= 7 \times 9$

_____ $= 8 \times 8$

3. Goals scored by players during a soccer season: 14, 12, 8, 7, 8, 11

The average (mean) number of goals is _____.

The number of goals that occurs most often (mode) is _____.

4. Is a 6-sided die more likely to land on an odd number or an even number? Explain.

____

____

____

____

____

5. Practice lattice multiplication.

$56 \times 78 =$ _________

6. Fill in the missing numbers on the number line.

_____ _____ −4 2 _____

Date Time

Spinners

Math Message

Color each circle so that it matches the description.

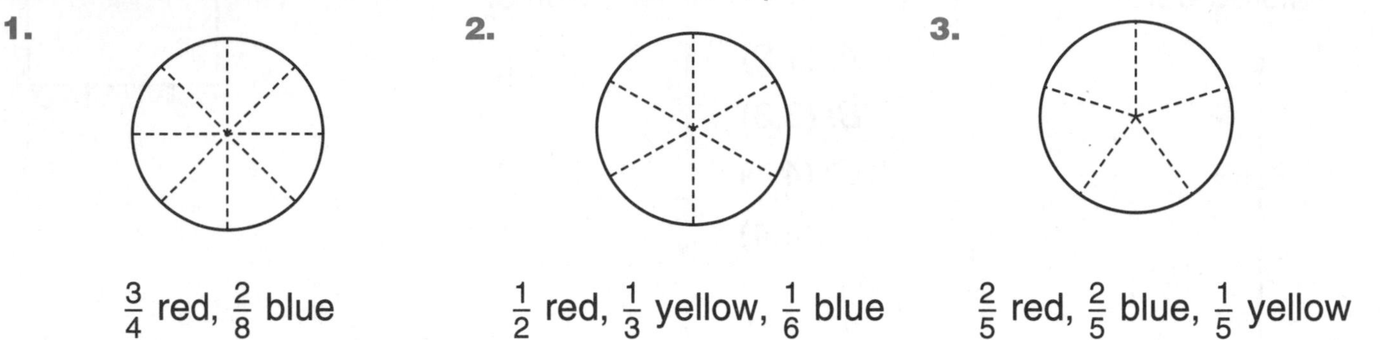

1. $\frac{3}{4}$ red, $\frac{2}{8}$ blue

2. $\frac{1}{2}$ red, $\frac{1}{3}$ yellow, $\frac{1}{6}$ blue

3. $\frac{2}{5}$ red, $\frac{2}{5}$ blue, $\frac{1}{5}$ yellow

---

Spinner Experiments

Tape *Math Masters,* page 180, to your desk or table. Make a spinner on the first circle.

4. Spin the paper clip 10 times. Tally the number of times the paper clip lands on the shaded part and on the white part.

Lands On	Tallies
shaded part	
white part	

5. Record results for the whole class.

Lands On	Totals
shaded part	
white part	

Make a spinner on the second circle.

6. Spin the paper clip 10 times. Tally the number of times the paper clip lands on the shaded part and on the white part.

Lands On	Tallies
shaded part	
white part	

7. Record results for the whole class.

Lands On	Totals
shaded part	
white part	

8. The paper clip has a better chance of landing on the _______________ part of the spinner than on the _______________ part.

Date Time

Degrees in a Turn

Tell how many degrees are in each turn.

1. full turn ________
2. half-turn ________
3. quarter-turn ________
4. $\frac{3}{4}$ turn ________
5. $\frac{1}{3}$ turn ________
6. $\frac{2}{3}$ turn ________
7. $\frac{1}{6}$ turn ________
8. $\frac{2}{6}$ turn ________
9. $\frac{3}{6}$ turn ________
10. $\frac{5}{6}$ turn ________

Challenge

Tell how many degrees are in each turn.

11. $\frac{1}{2}$ of a half-turn ________
12. $\frac{1}{2}$ of a quarter-turn ________
13. $\frac{1}{3}$ of a half-turn ________
14. $\frac{1}{3}$ of a quarter-turn ________

Tell what fraction of a full turn each angle is.

15. 30° _____
16. 45° _____

Date Time

Math Boxes 11.4

1. Write 5 fractions that are greater than $\frac{9}{12}$.

_____ _____ _____ _____ _____

Write 5 fractions that are less than $\frac{9}{12}$.

_____ _____ _____ _____ _____

Write 3 other names for $\frac{9}{12}$.

_____ _____ _____

SRB 30

2. Complete.

2 quarts = ______ pints

______ quarts = 30 pints

2 quarts = ______ fl oz

______ quarts = 96 fl oz

2 gallons = ______ fl oz

SRB 142 143

3. Complete.

Area of Base (square inches)	Height (inches)	Volume (cubic inches)
100		900
50		3,500
20		1,800
70		4,200

SRB 139–141

4. Write the number that has

0 in the thousandths place

4 in the ones place

1 in the tenths place

8 in the hundredths place

___ . ___ ___ ___

5. Describe 2 events that have a "good chance" of happening.

____

____

____

____

____

SRB 84–86

6. A van holds 8 people. 125 people are going to the concert. How many vans are needed?

_____ vans

Will all of the vans be full? _____

Date Time

Math Boxes 11.5

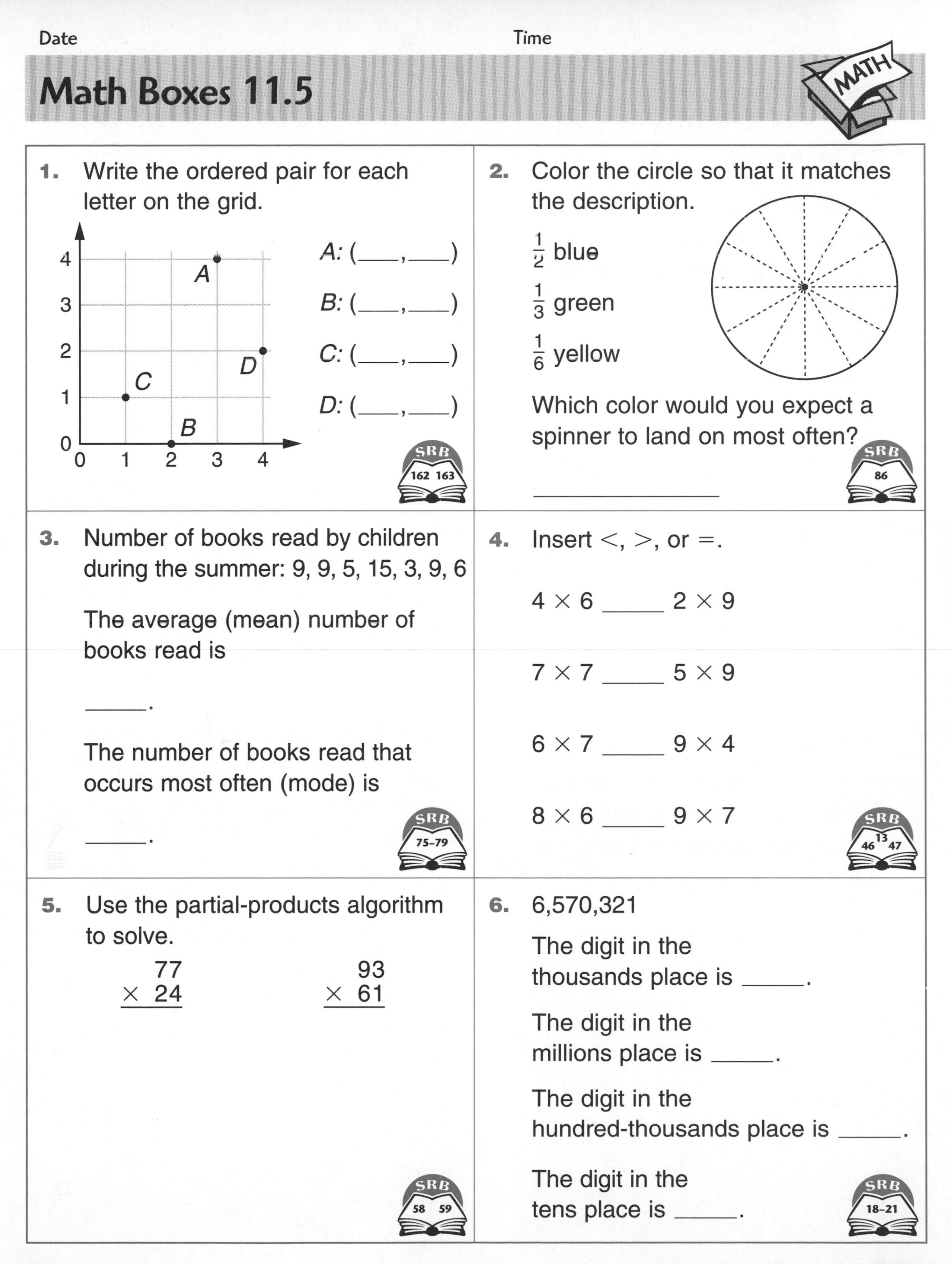

1. Write the ordered pair for each letter on the grid.

A: (___,___)

B: (___,___)

C: (___,___)

D: (___,___)

SRB 162 163

2. Color the circle so that it matches the description.

$\frac{1}{2}$ blue

$\frac{1}{3}$ green

$\frac{1}{6}$ yellow

Which color would you expect a spinner to land on most often?

____

SRB 86

3. Number of books read by children during the summer: 9, 9, 5, 15, 3, 9, 6

The average (mean) number of books read is

_____.

The number of books read that occurs most often (mode) is

_____.

SRB 75–79

4. Insert <, >, or =.

4×6 _____ 2×9

7×7 _____ 5×9

6×7 _____ 9×4

8×6 _____ 9×7

SRB 13 46 47

5. Use the partial-products algorithm to solve.

$$\begin{array}{r} 77 \\ \times\ 24 \\ \hline \end{array} \qquad \begin{array}{r} 93 \\ \times\ 61 \\ \hline \end{array}$$

SRB 58 59

6. 6,570,321

The digit in the thousands place is _____.

The digit in the millions place is _____.

The digit in the hundred-thousands place is _____.

The digit in the tens place is _____.

SRB 18–21

Date Time

Making Spinners

Math Message

1. Use exactly six different colors. Make a spinner so that the paper clip has the **same chance** of landing on any one of the six colors.

   (Hint: Into how many equal parts should the circle be divided?)

360°
0°
345° 15°
330° 30°
315° 45°
300° 60°
285° 75°
270° 90°
255° 105°
240° 120°
225° 135°
210° 150°
195° 180° 165°

2. Use only blue and red. Make a spinner so that the paper clip is **twice as likely** to land on blue as it is to land on red.

360°
0°
345° 15°
330° 30°
315° 45°
300° 60°
285° 75°
270° 90°
255° 105°
240° 120°
225° 135°
210° 150°
195° 180° 165°

Date Time

Making Spinners (cont.)

3. Use only blue, red, and green. Make a spinner so that the paper clip:

- has the **same chance** of landing on blue and on red

and

- is **less likely** to land on green than on blue.

360° 0° 15° 30° 45° 60° 75° 90° 105° 120° 135° 150° 165° 180° 195° 210° 225° 240° 255° 270° 285° 300° 315° 330° 345°

4. Use only blue, red, and yellow. Make a spinner so that the paper clip:

- is **more likely** to land on blue than on red

and

- is **less likely** to land on yellow than on blue.

360° 0° 15° 30° 45° 60° 75° 90° 105° 120° 135° 150° 165° 180° 195° 210° 225° 240° 255° 270° 285° 300° 315° 330° 345°

Date Time

Parentheses Puzzles

Circle the expressions that belong in each name-collection box.

1. 7

$7 \times (8 - 7)$ $(5 + 5) - (3 - 1)$

$(3 \times 4) - 5$ $(3 \times 7) \div 7$

$(12 + 2) \div 2$ $(4 \times 7) \div (10 - 6)$

$(7 \times 1) \times 7$

2. 25

$(4 + 1) \times (6 - 1)$ $100 \div (6 - 1)$

$(4 \times 5) + 5$ $(2 \times 10) \times 5$

$25 + (0 \times 10)$ $50 - (30 - 10)$

$(6 \times 5) - (1 \times 5)$

3. Fill in the tag.
Circle correct names.
Cross out incorrect names.
Add 2 names.

$5 + (10 \div 2)$ $9 + (10 \div 10)$

$(3 + 2) \times 2$ $(9 + 10) \div 10$

$(20 \div 4) + 5$ $4 + (2 \times 3)$

$2 + (4 \times 2)$

$(2 + 4) \times 2$

Add parentheses, ( ), to complete the number models.

4. $7 \times 8 - 8 = 48$

5. $7 \times 8 - 8 = 0$

6. $280 = 85 - 45 \times 22 - 15$

7. $80 \times 3 + 4 = 560$

Complete the number models.

8. ______ $= (24 \div 6) - 2$

9. ______ $= 24 \div (6 - 2)$

10. $(7 + 3) \times (9 + 6) =$ ______

11. $7 + (3 \times 9) + 6 =$ ______

Date Time

Drawing Blocks

Color the blocks in the bags blue. Then fill in the blanks by answering this question: How many red blocks would you put into each bag?

1. If I wanted to have an equal chance of taking out red or blue, I would put in _____ red block(s).

2. If I wanted to be more likely to take out blue than red, I would put in _____ red block(s).

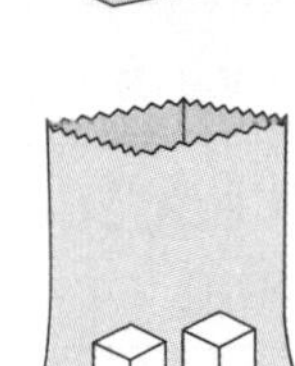

3. If I wanted to be sure of taking out a blue block, I would put in _____ red block(s).

4. If I wanted to take out a red block about 3 times as often as a blue one, I would put in _____ red block(s).

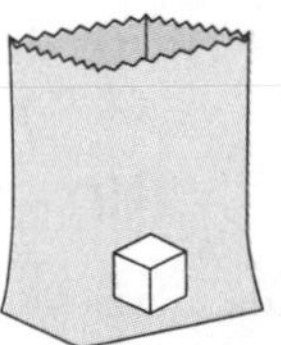

5. If I wanted to take out a red block about half as often as a blue one, I would put in _____ red block(s).

6. If I wanted to take out a red block about $\frac{1}{3}$ of the time, I would put in _____ red block(s).

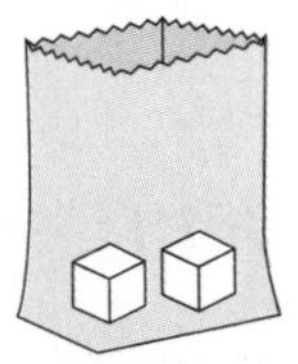

Challenge

7. If I wanted to take out a red block about $\frac{2}{3}$ of the time, I would put in _____ red block(s).

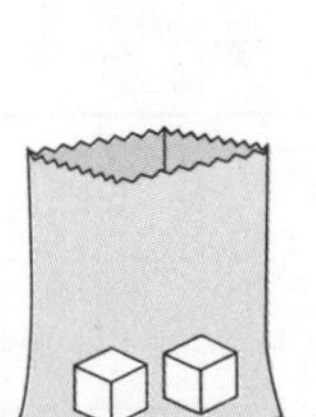

Date Time

The Best Pizza

Pizza Prices	10-inch serves 2–3	12-inch serves 3–4	14-inch serves 4–5
plain cheese	$8.35	$10.85	$12.40
each added topping	$1.25	$1.50	$1.75

Choices for added toppings: sausage, mushrooms, red or green peppers, onions, pepperoni, spinach, ground beef, extra cheese

1. How much does a 10-inch pizza with 1 added topping cost? $________

2. How much does a 12-inch pizza with 2 added toppings cost? $________

3. For the class pizza party, the class votes to order five 14-inch pizzas with one added topping on each pizza. How much will the pizzas cost? $________

4. Which costs more? Circle your answer.

   3 of the 10-inch pizzas with 2 added toppings each

   2 of the 14-inch pizzas with 1 added topping each

5. Write and solve your own pizza number story.

____

____

____

____

Date Time

Math Boxes 11.6

1. Find the volume of the rectangular prism.

Volume = _____ cubic units

SRB 139–141

2. Write six numbers that are factors of 100.

_____ _____ _____

_____ _____ _____

SRB 37

3. A large bag of candy costs $3.59. What is the cost of 6 bags?

Fill in the oval next to the best estimate.

0 $15.00

0 $18.00

0 $21.00

0 $24.00

SRB 167 168

4. Put the fractions in order from smallest to largest.

$\frac{1}{5}$, $\frac{1}{7}$, $\frac{3}{4}$, $\frac{5}{6}$, $\frac{99}{100}$

_____, _____, _____, _____, _____

SRB 31 32

5. Describe 2 events that are "not likely" to happen.

____

____

____

____

____

SRB 84

6. Draw a shape with an area of 10 square centimeters.

What is the perimeter of your shape? _____centimeters

SRB 132 133 136–138

Date Time

Random-Draw Problems

Each problem is about marbles in a jar. The marbles may be black, white, or striped. A marble is drawn at random from the jar (without looking into the jar). The kind of marble is tallied. Then the marble is returned to the jar.

- Decide, from the description of the random draws in each problem, how many marbles of each kind are in the jar.
- Shade the circles in the jar to match your decision.

1. From 100 random draws, you get:

a black marble ● 62 times

a white marble ○ 38 times

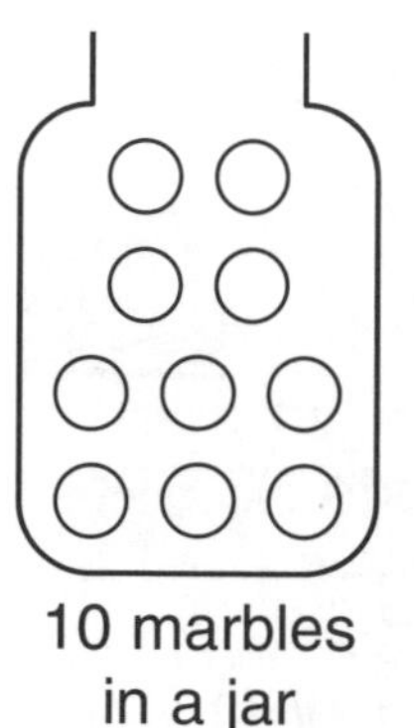
10 marbles in a jar

2. From 50 random draws, you get:

a black marble ● 30 times

a white marble ○ 16 times

a striped marble ⦶ 4 times

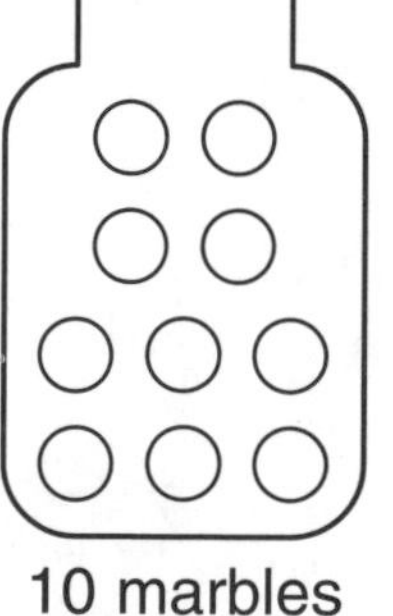
10 marbles in a jar

3. From 100 random draws, you get:

a black marble ● 23 times

a white marble ○ 53 times

a striped marble ⦶ 24 times

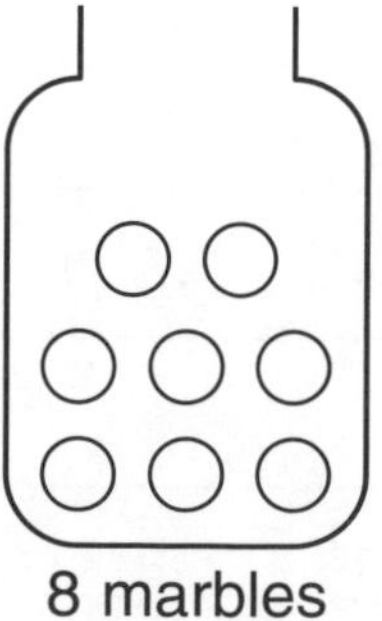
8 marbles in a jar

Date Time

Math Boxes 11.7

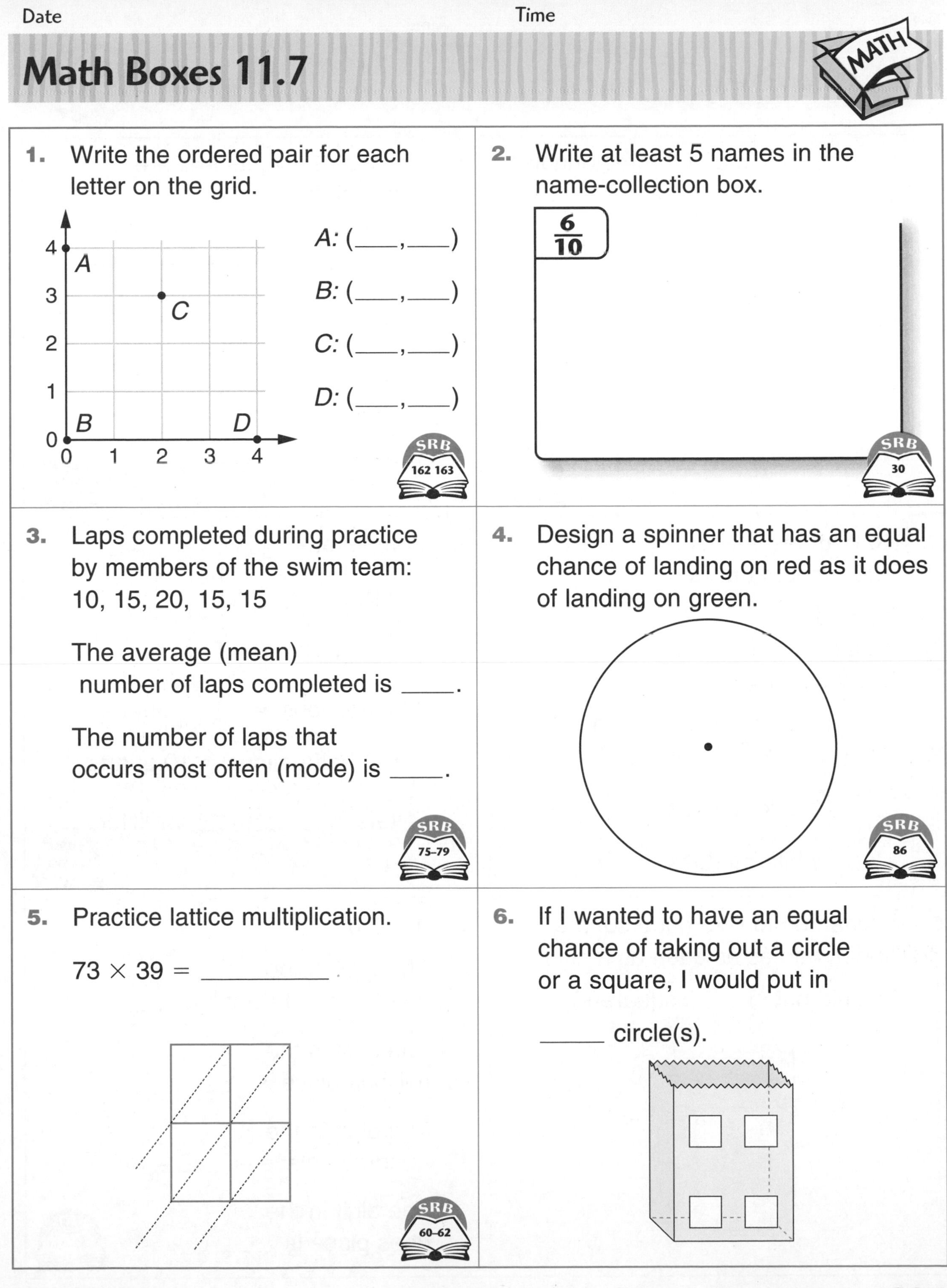

1. Write the ordered pair for each letter on the grid.

A: (___,___)

B: (___,___)

C: (___,___)

D: (___,___)

SRB 162 163

2. Write at least 5 names in the name-collection box.

$\frac{6}{10}$

SRB 30

3. Laps completed during practice by members of the swim team: 10, 15, 20, 15, 15

The average (mean) number of laps completed is ____.

The number of laps that occurs most often (mode) is ____.

SRB 75–79

4. Design a spinner that has an equal chance of landing on red as it does of landing on green.

SRB 86

5. Practice lattice multiplication.

$73 \times 39 =$ __________

SRB 60–62

6. If I wanted to have an equal chance of taking out a circle or a square, I would put in

_____ circle(s).

Date Time

Math Boxes 11.8

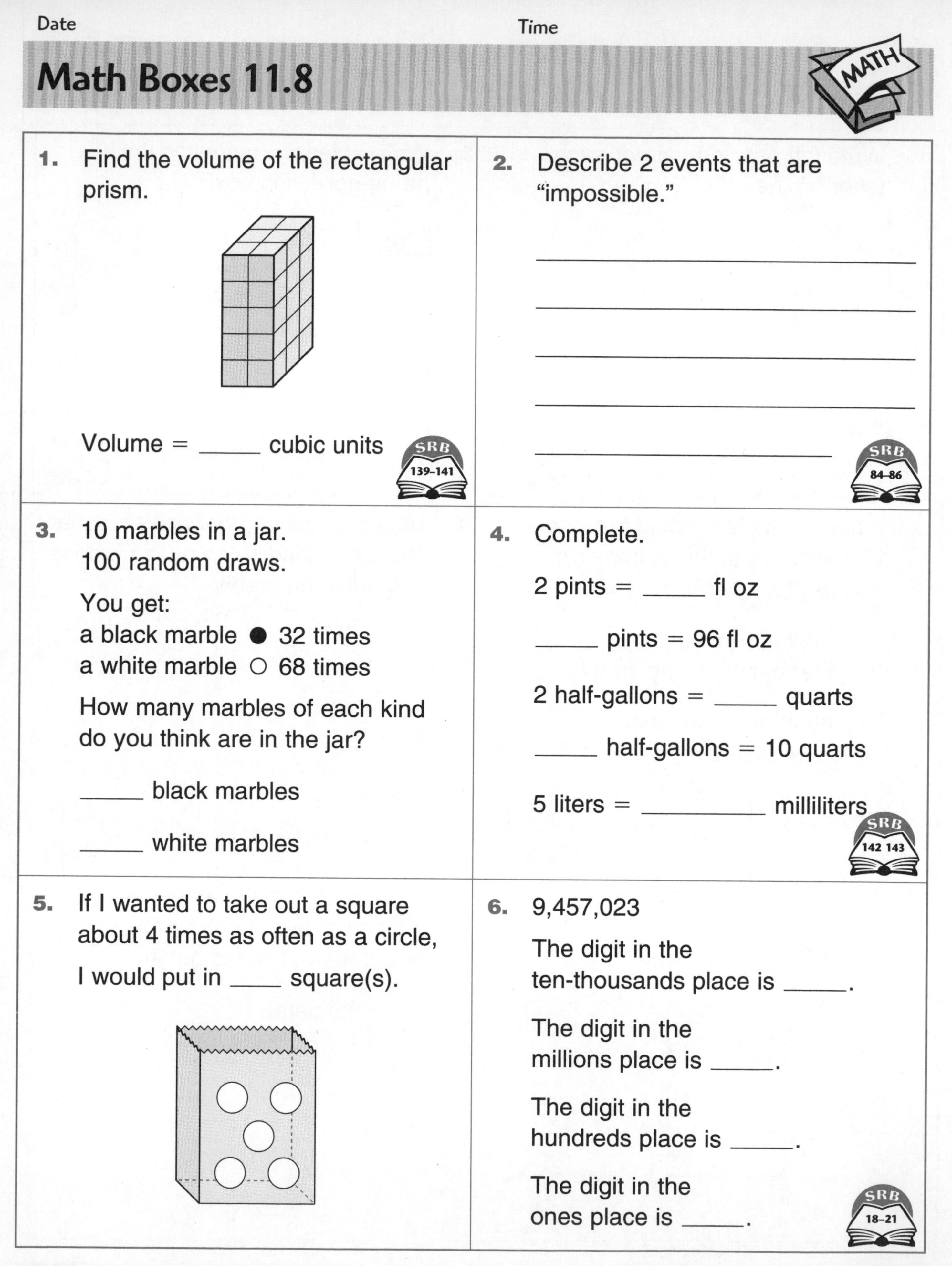

1. Find the volume of the rectangular prism.

Volume = _____ cubic units

SRB 139–141

2. Describe 2 events that are "impossible."

____

____

____

____

____

SRB 84–86

3. 10 marbles in a jar.
100 random draws.

You get:
a black marble ● 32 times
a white marble ○ 68 times

How many marbles of each kind do you think are in the jar?

_____ black marbles

_____ white marbles

4. Complete.

2 pints = _____ fl oz

_____ pints = 96 fl oz

2 half-gallons = _____ quarts

_____ half-gallons = 10 quarts

5 liters = _________ milliliters

SRB 142 143

5. If I wanted to take out a square about 4 times as often as a circle,

I would put in ____ square(s).

6. 9,457,023

The digit in the ten-thousands place is _____.

The digit in the millions place is _____.

The digit in the hundreds place is _____.

The digit in the ones place is _____.

SRB 18–21

Date Time

Estimate—Then Calculate

Solve only the problems whose sum or difference is greater than 500.

1.

$$\begin{array}{r} 825 \\ -\ 347 \\ \hline \end{array}$$

2.

$$\begin{array}{r} 984 \\ -\ 392 \\ \hline \end{array}$$

3.

$$\begin{array}{r} 658 \\ -\ 179 \\ \hline \end{array}$$

4.

$$\begin{array}{r} 324 \\ +\ 161 \\ \hline \end{array}$$

5.

$$\begin{array}{r} 728 \\ -\ 232 \\ \hline \end{array}$$

6.

$$\begin{array}{r} 227 \\ +\ 285 \\ \hline \end{array}$$

7.

$$\begin{array}{r} 174 \\ +\ 338 \\ \hline \end{array}$$

8.

$$\begin{array}{r} 881 \\ -\ 293 \\ \hline \end{array}$$

9.

$$\begin{array}{r} 1{,}294 \\ -\ 776 \\ \hline \end{array}$$

Date Time

Math Boxes 11.9

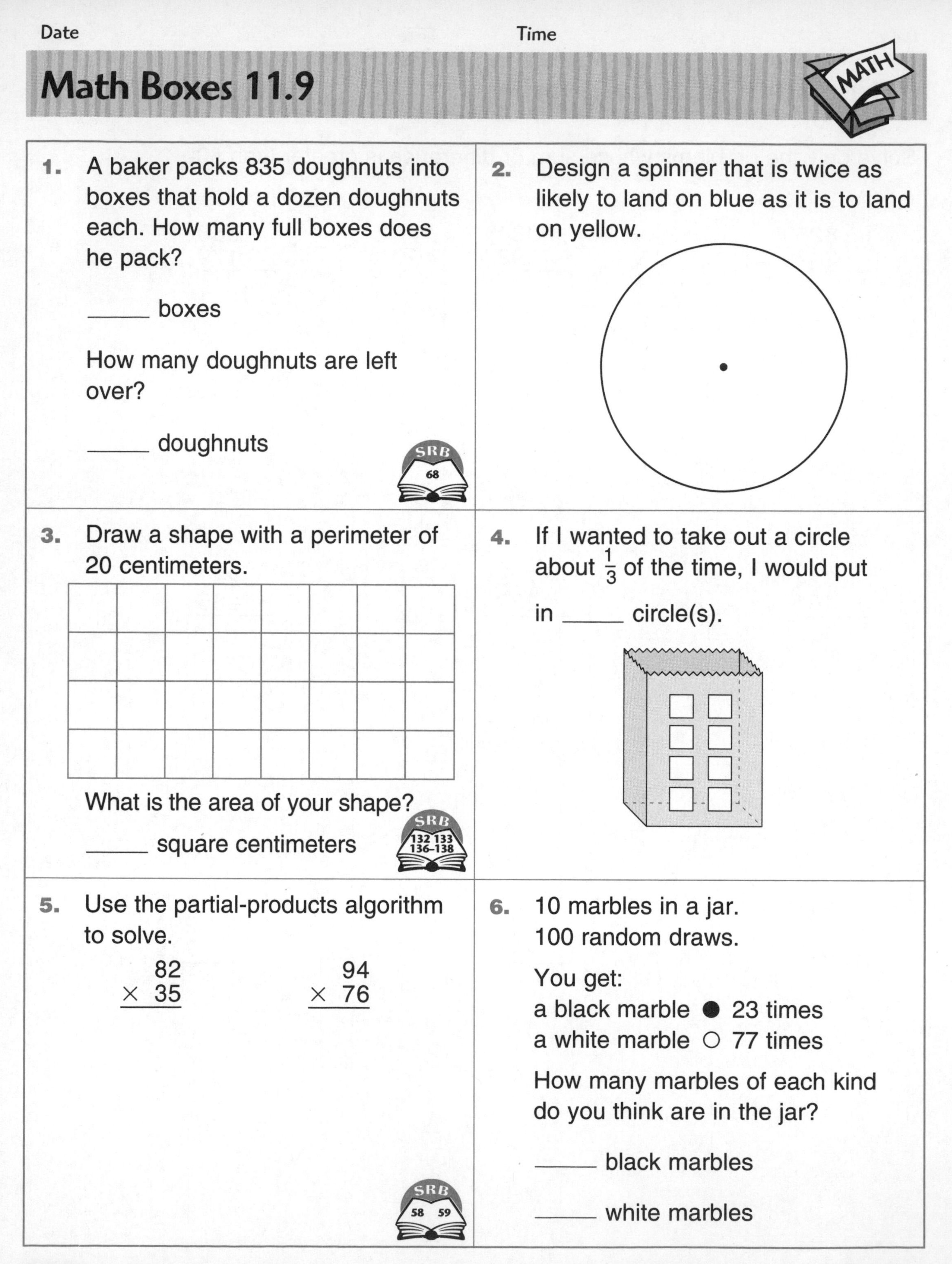

1. A baker packs 835 doughnuts into boxes that hold a dozen doughnuts each. How many full boxes does he pack?

   _____ boxes

   How many doughnuts are left over?

   _____ doughnuts

   SRB 68

2. Design a spinner that is twice as likely to land on blue as it is to land on yellow.

3. Draw a shape with a perimeter of 20 centimeters.

   What is the area of your shape?

   _____ square centimeters

   SRB 132 133 136–138

4. If I wanted to take out a circle about $\frac{1}{3}$ of the time, I would put in _____ circle(s).

5. Use the partial-products algorithm to solve.

   $$\begin{array}{r} 82 \\ \times\ 35 \\ \hline \end{array} \qquad \begin{array}{r} 94 \\ \times\ 76 \\ \hline \end{array}$$

   SRB 58 59

6. 10 marbles in a jar.
   100 random draws.

   You get:
   a black marble ● 23 times
   a white marble ○ 77 times

   How many marbles of each kind do you think are in the jar?

   _____ black marbles

   _____ white marbles

Math Boxes 11.10

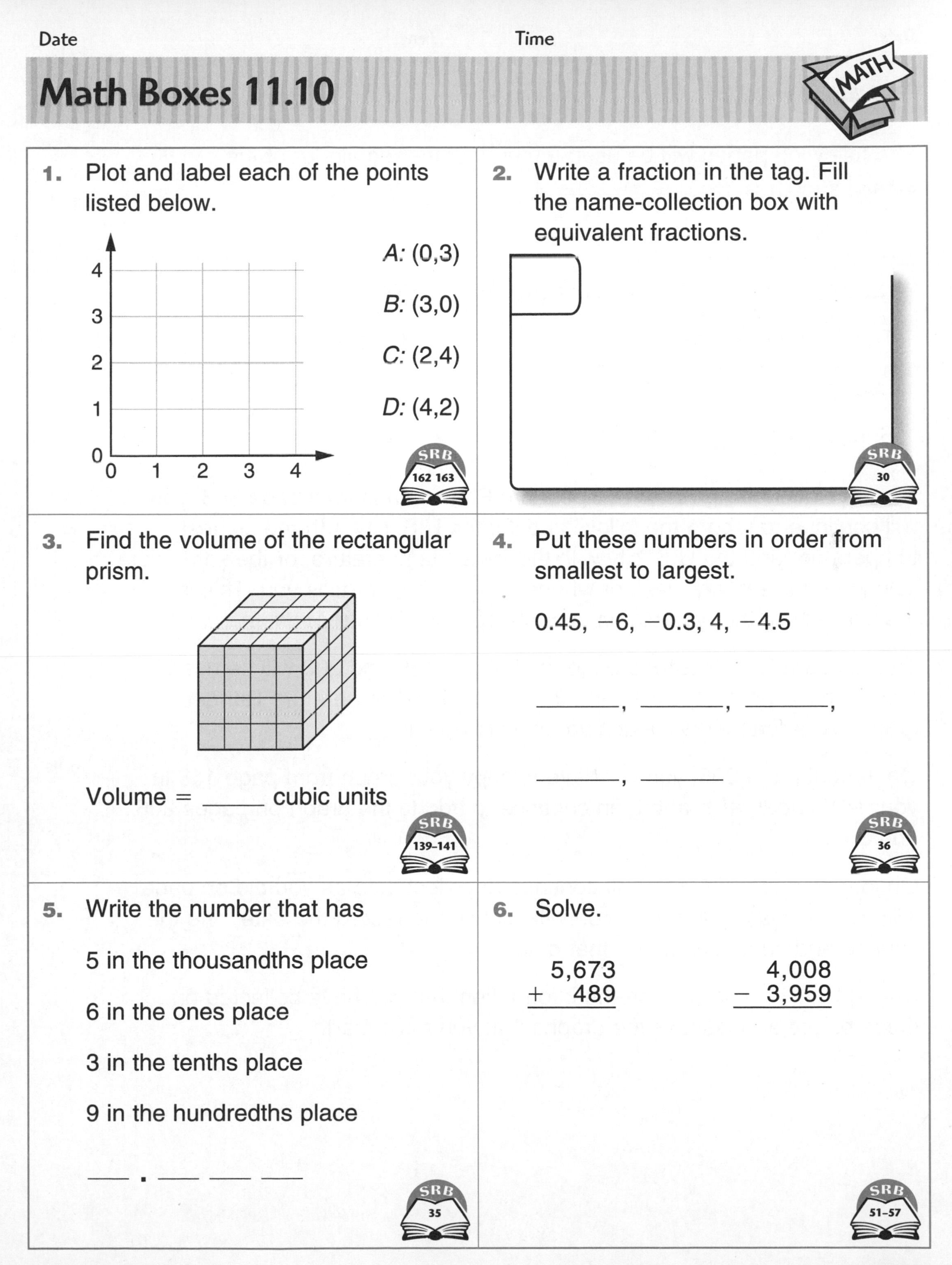

1. Plot and label each of the points listed below.

   A: (0,3)

   B: (3,0)

   C: (2,4)

   D: (4,2)

   SRB 162 163

2. Write a fraction in the tag. Fill the name-collection box with equivalent fractions.

   SRB 30

3. Find the volume of the rectangular prism.

   Volume = _____ cubic units

   SRB 139–141

4. Put these numbers in order from smallest to largest.

   0.45, −6, −0.3, 4, −4.5

   ______, ______, ______,

   ______, ______

   SRB 36

5. Write the number that has

   5 in the thousandths place

   6 in the ones place

   3 in the tenths place

   9 in the hundredths place

   ___ . ___ ___ ___

   SRB 35

6. Solve.

$$\begin{array}{r} 5{,}673 \\ +\ \ 489 \\ \hline \end{array} \qquad \begin{array}{r} 4{,}008 \\ -\ 3{,}959 \\ \hline \end{array}$$

   SRB 51–57

Date Time

Special Pages

The following pages will be used throughout the remainder of the school year.

	Pages
National High/Low Temperature Project	303
Temperature Ranges	304 and 305
Length of Day	306 and 307
Sunrise and Sunset Record	308

For the National High/Low Temperature Project on journal page 303, you will continue to record the following data: the U.S. city with the highest temperature and the U.S. city with the lowest temperature for the same date. You will do this every week or whenever your teacher tells you. This is the same thing that you did on journal page 160 in your *Math Journal 1.*

The data that you recorded on journal pages 160 and 303 will be used in Unit 7 on journal pages 304 and 305 to make a Temperature Ranges graph. Your teacher will teach you how to do this.

On journal page 306, you will have to copy your graph from page 158 in your *Math Journal 1,* and then continue to add to the graph on pages 306 and 307.

On journal page 308, you will continue to collect data as you did on page 158 in your *Math Journal 1.* You will continue to record the date, and the sunrise and sunset times for that date.

During Unit 11, you will use the information that you have collected on these pages and discuss the graphs that you have made.

Date Time

National High/Low Temperatures Project

Date	Highest Temperature		Lowest Temperature		Difference in Temperature
	Place	Temperature	Place	Temperature	
		°F		°F	°F
		°F		°F	°F
		°F		°F	°F
		°F		°F	°F
		°F		°F	°F
		°F		°F	°F
		°F		°F	°F
		°F		°F	°F
		°F		°F	°F
		°F		°F	°F
		°F		°F	°F
		°F		°F	°F
		°F		°F	°F
		°F		°F	°F
		°F		°F	°F
		°F		°F	°F
		°F		°F	°F
		°F		°F	°F
		°F		°F	°F
		°F		°F	°F
		°F		°F	°F

Date Time

Temperature Ranges

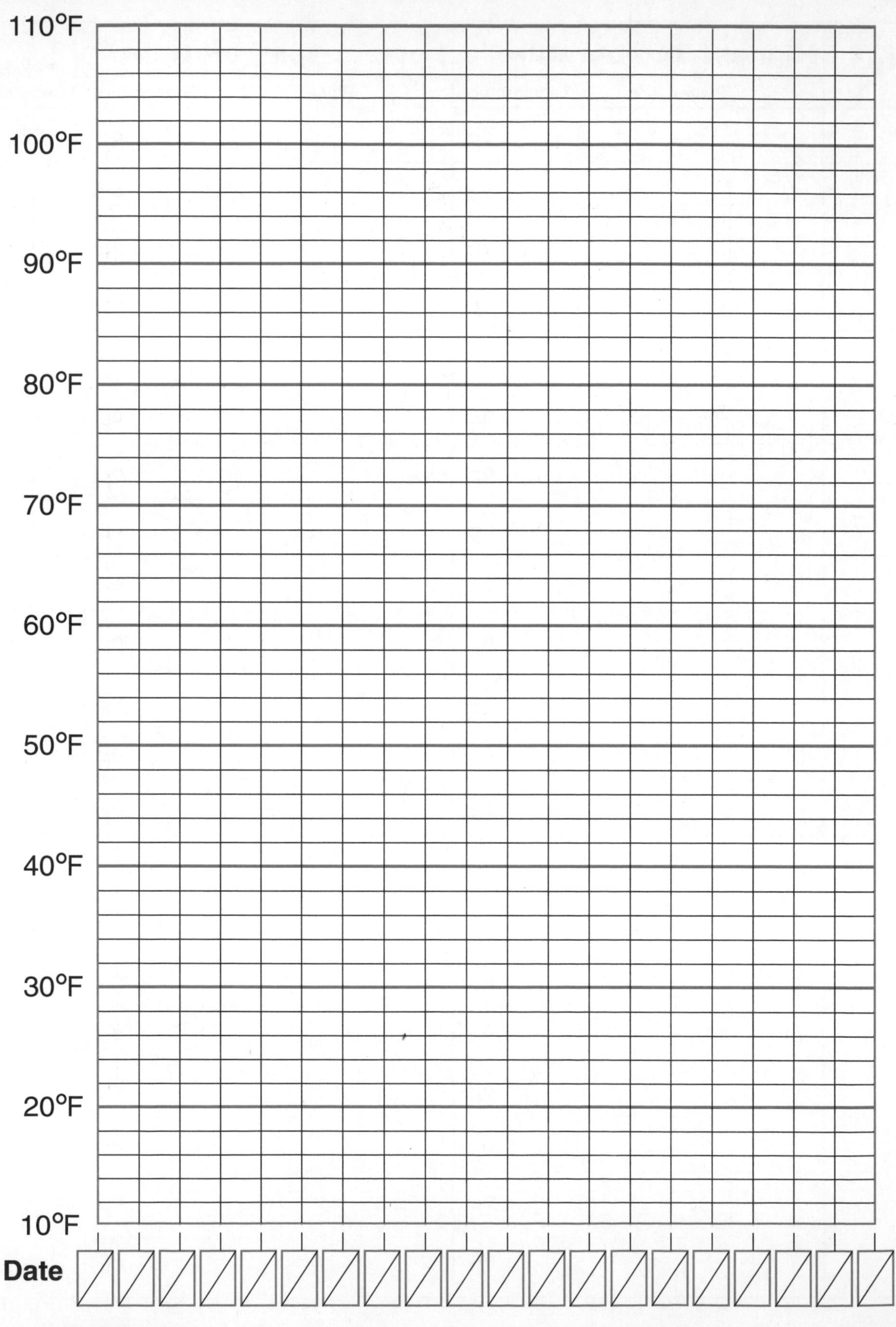

 Use with Lesson 7.8.

Date Time

Temperature Ranges (cont.)

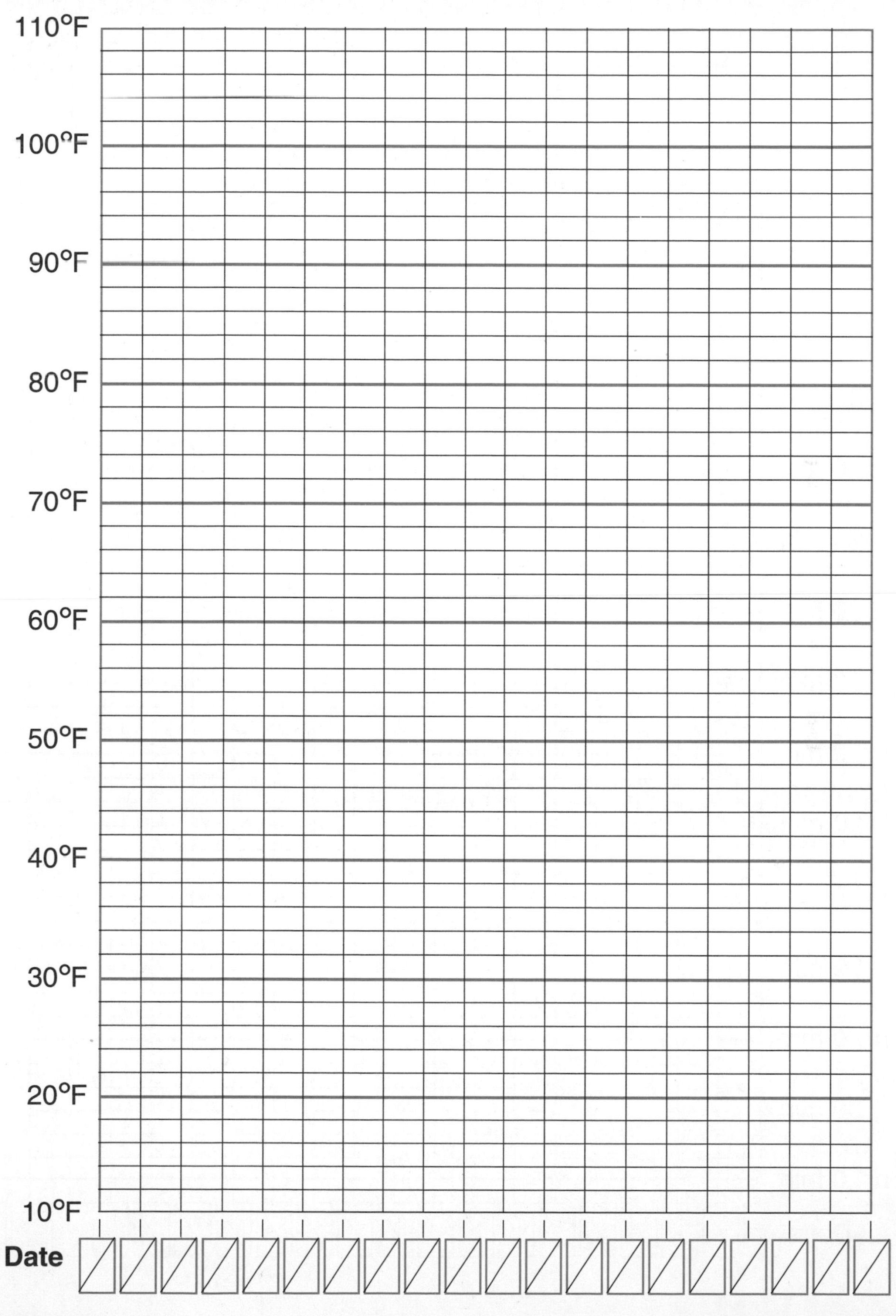

Date Time

Length of Day

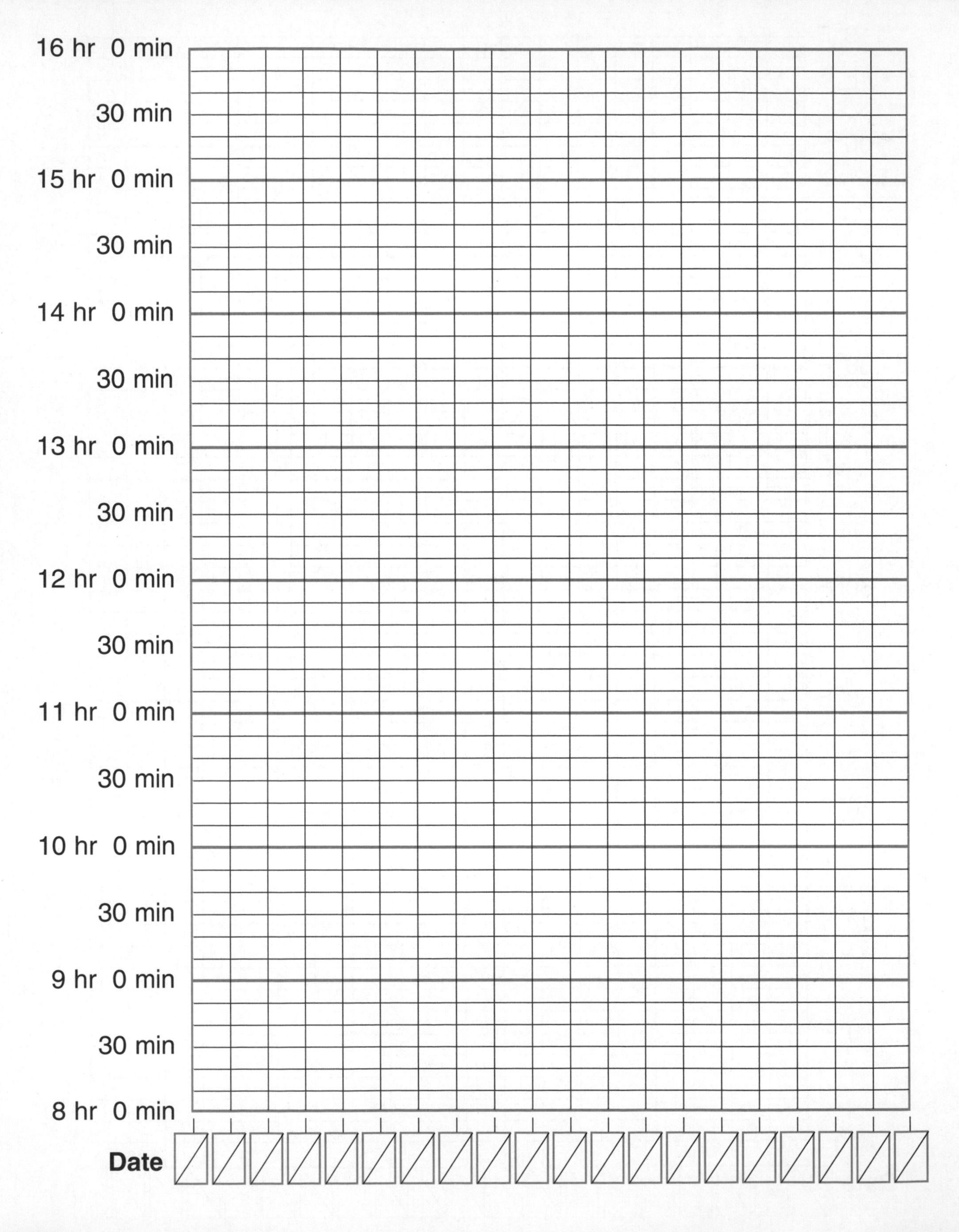

Date Time

Length of Day

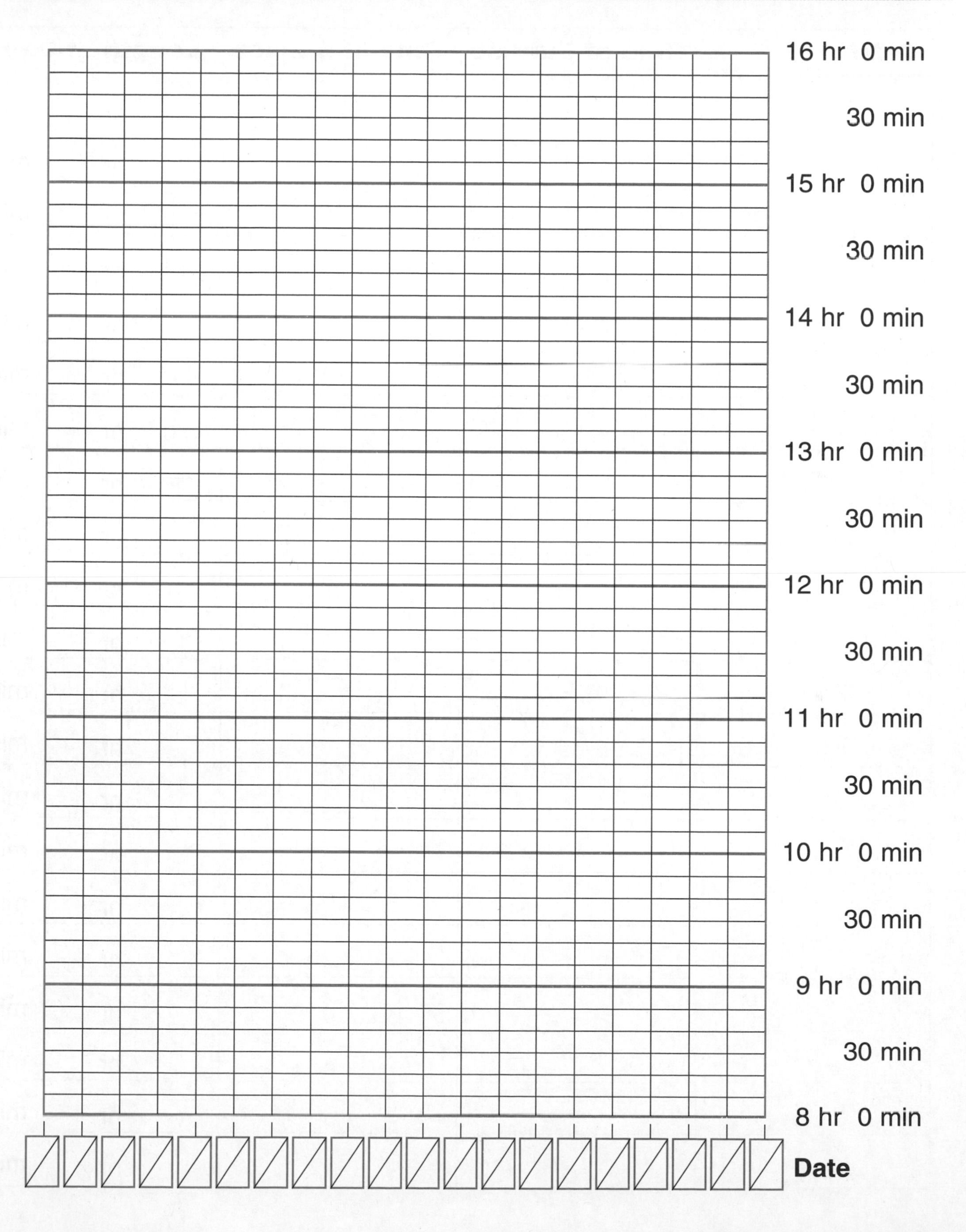

Date Time

Sunrise and Sunset Record

Date	Time of Sunrise	Time of Sunset	Length of Day
			hr min
			hr min
			hr min
			hr min
			hr min
			hr min
			hr min
			hr min
			hr min
			hr min
			hr min
			hr min
			hr min
			hr min
			hr min
			hr min
			hr min
			hr min
			hr min
			hr min
			hr min

Use with Lesson 7.8.

Date Time

Notes

Date Time

Notes

Date Time

Notes

Date Time

Notes

Date Time

×, ÷ Fact Triangles 3

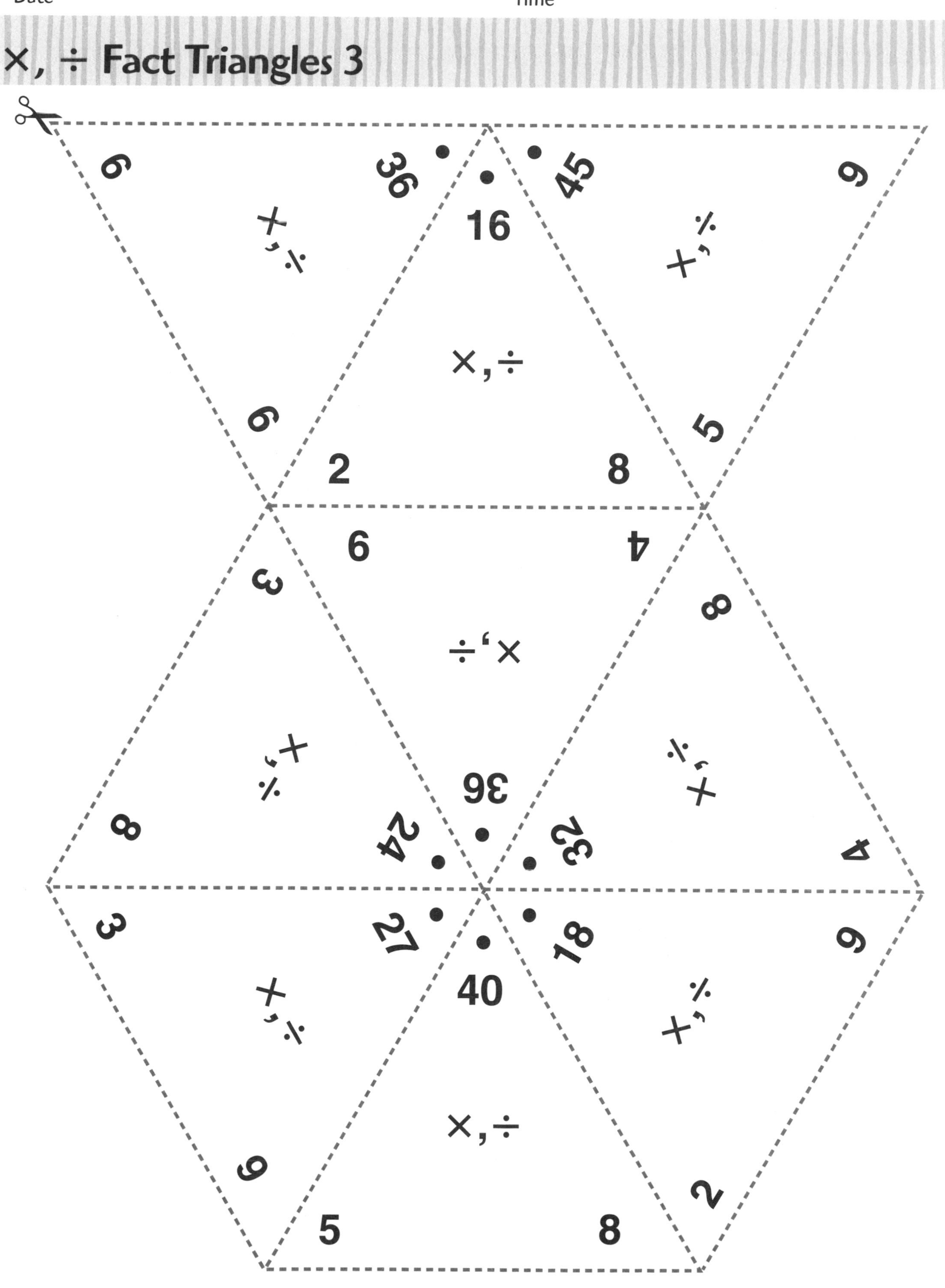

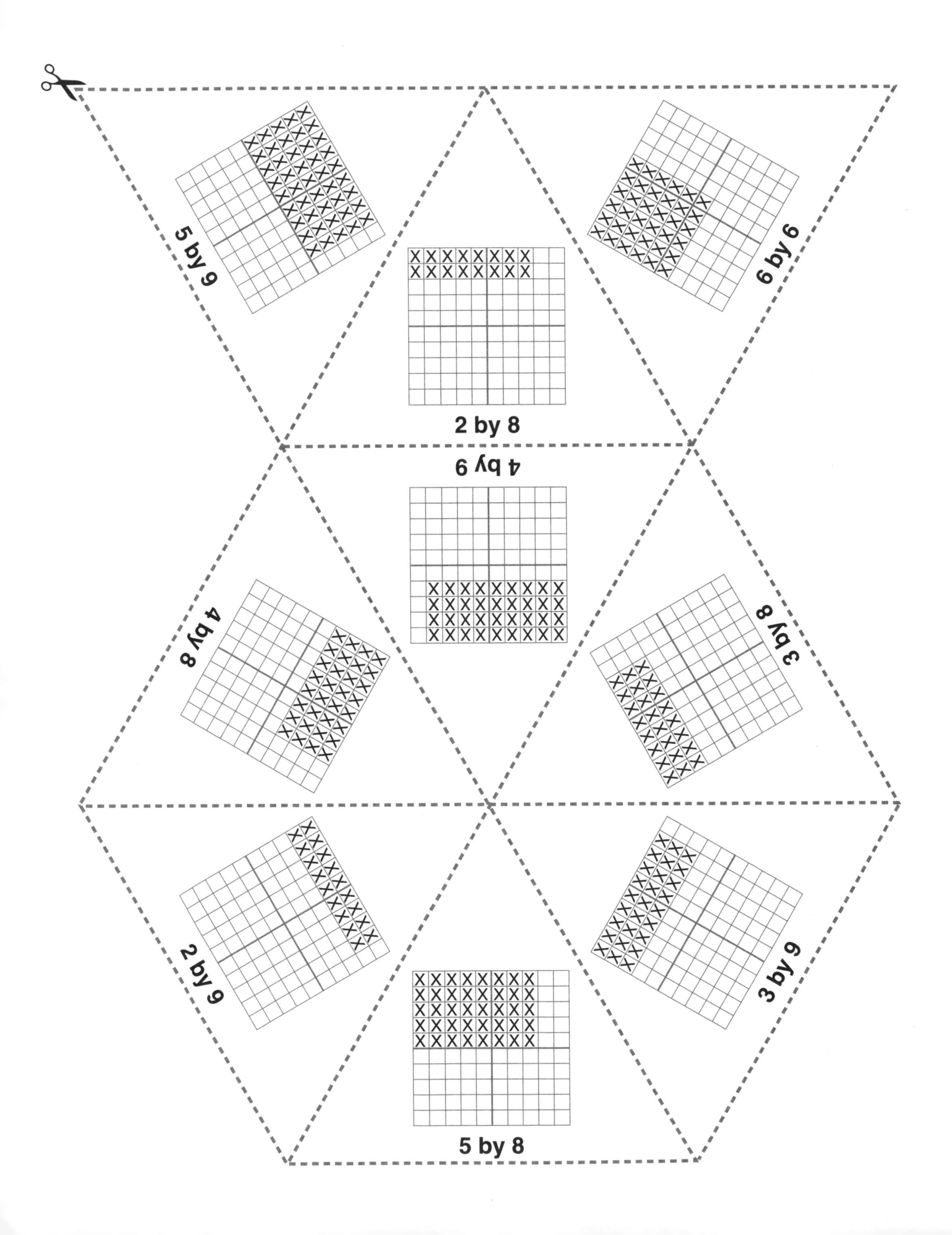
5 by 9
2 by 8
6 by 6
4 by 9
4 by 8
3 by 8
2 by 9
5 by 8
3 by 9

Date Time

×, ÷ Fact Triangles 4

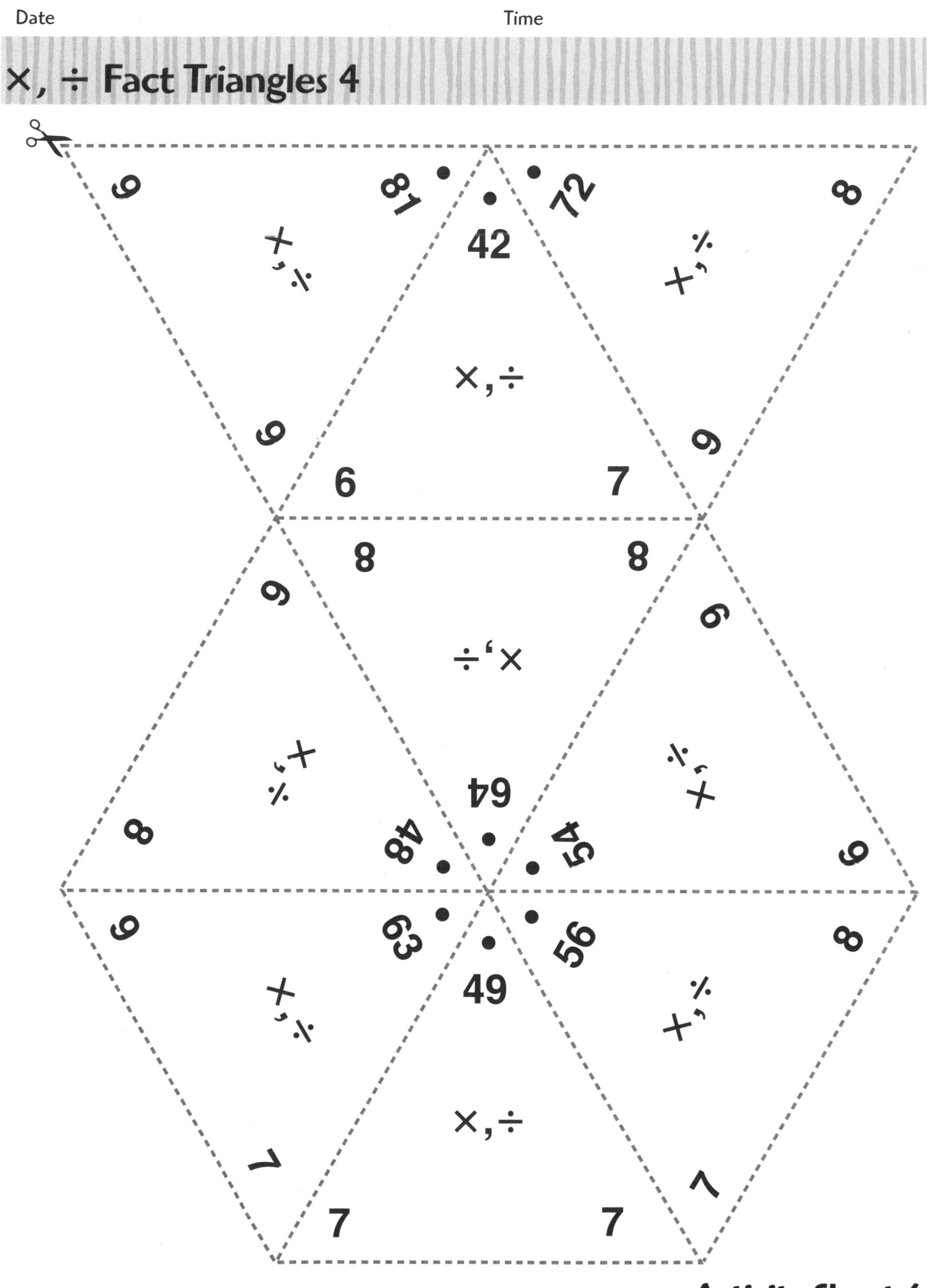

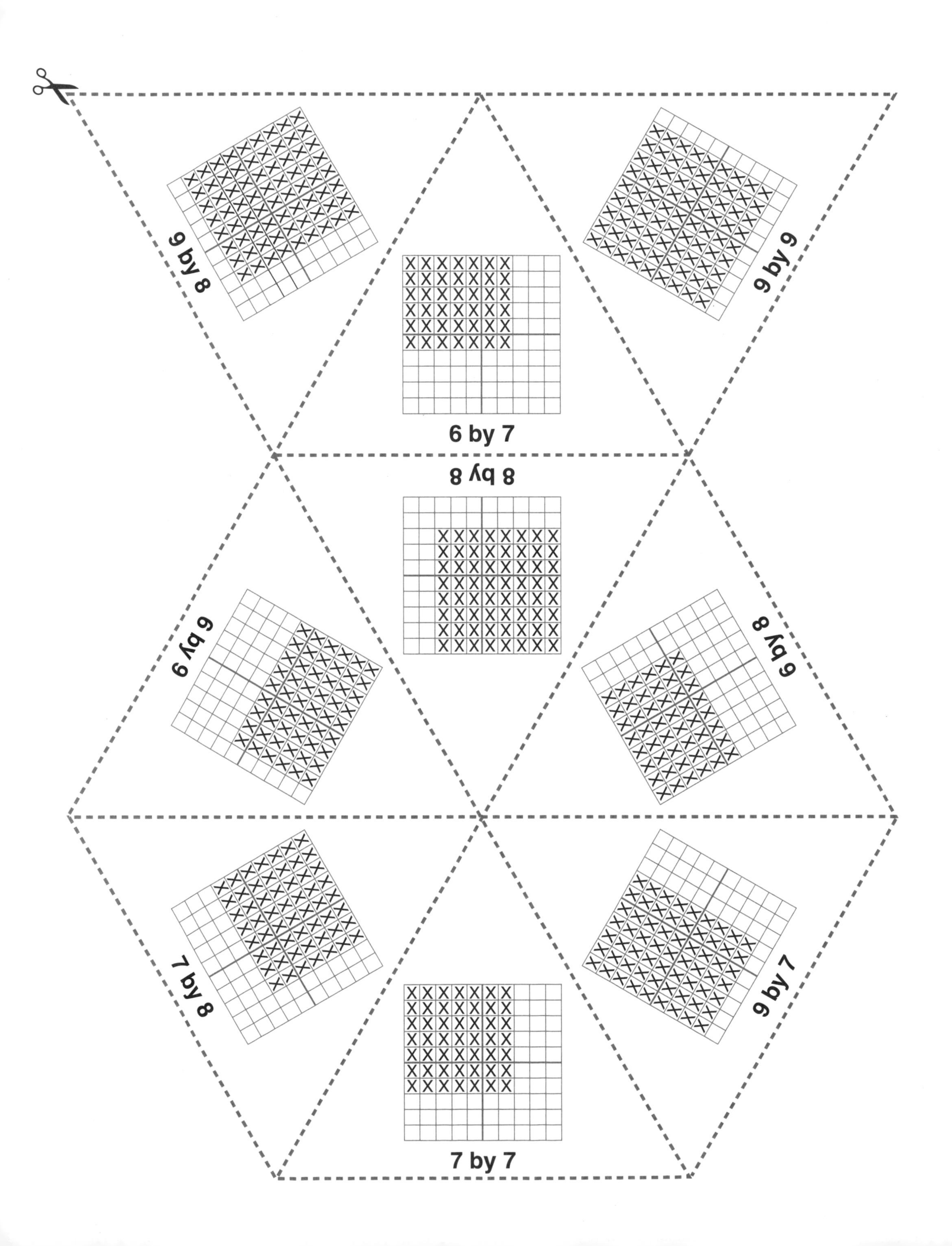

8 by 9
6 by 7
9 by 9
8 by 8
6 by 9
6 by 8
7 by 8
7 by 7
9 by 7

Date Time

Fraction Cards

$\frac{1}{2}$	$\frac{3}{6}$	$\frac{1}{3}$	$\frac{2}{2}$
$\frac{2}{6}$	$\frac{1}{4}$	$\frac{2}{3}$	$\frac{3}{4}$
$\frac{2}{4}$	$\frac{6}{8}$	$\frac{0}{2}$	$\frac{4}{6}$
$\frac{4}{8}$	$\frac{0}{4}$	$\frac{2}{8}$	$\frac{4}{4}$

$\frac{2}{2}$	$\frac{1}{3}$	$\frac{3}{6}$	$\frac{1}{2}$
$\frac{3}{4}$	$\frac{2}{3}$	$\frac{1}{4}$	$\frac{2}{6}$
$\frac{4}{6}$	$\frac{0}{2}$	$\frac{6}{8}$	$\frac{2}{4}$
$\frac{4}{4}$	$\frac{2}{8}$	$\frac{0}{4}$	$\frac{4}{8}$

Date Time

Fraction Cards

$\frac{5}{10}$ $\frac{8}{12}$ $\frac{3}{9}$ $\frac{6}{12}$

$\frac{1}{6}$ $\frac{9}{12}$ $\frac{5}{6}$ $\frac{2}{12}$

$\frac{1}{5}$ $\frac{6}{9}$ $\frac{3}{12}$ $\frac{2}{10}$

$\frac{4}{12}$ $\frac{4}{5}$ $\frac{10}{12}$ $\frac{8}{10}$

$\frac{6}{12}$	$\frac{3}{9}$	$\frac{8}{12}$	$\frac{5}{10}$
$\frac{2}{12}$	$\frac{5}{6}$	$\frac{9}{12}$	$\frac{1}{6}$
$\frac{2}{10}$	$\frac{3}{12}$	$\frac{6}{9}$	$\frac{1}{5}$
$\frac{8}{10}$	$\frac{10}{12}$	$\frac{4}{5}$	$\frac{4}{12}$